Materials and Chemical
CHEMISTRY

Materials and Chemical CHEMISTRY

Barry Griffin

Oxford Book Company
Jaipur ▪ New Delhi

Published By

Oxford Book Company

Regd. Off. "Bony Residency", Gate No. 2
Opp. Tilak Public School, Vishweswariya Nagar,
Gopalpura Road, Jaipur - 302018, India
Ph: 0141 - 2761280, Telefax: 2761381
email - oxfordbookjpr@gmail.com
website - oxfordbookcompany.com

Delhi Office: 102, 1st Floor, Satyam House,
4327/3, Ansari Road, Darya Ganj,
New Delhi - 110002, India
Ph: 011-45652440

Distributed by
KEYA INTERNATIONAL PRESS LLP
E-mail: keyainternationalpressllp@gmail.com

First published in India, 2024

ISBN : 9789355240613

Typeset by Shivangi Computers, Jaipur

Printed and bound in India by Replika Press Pvt. Ltd.

Contents

Preface

Chemical engineering largely involves the design, improvement and maintenance of processes involving chemical or biological transformations for large-scale manufacture. Chemical engineers ensure the processes are operated safely, sustainably and economically. Chemical engineers in this branch are usually employed under the title of process engineer. A related term with a wider definition is chemical technology. In 1824, French physicist Sadi Carnot, in his "On the Motive Power of Fire", was the first to study the thermodynamics of combustion reactions in steam engines. In the 1850s, German physicist Rudolf Clausius began to apply the principles developed by Carnot to chemical systems at the atomic to molecular scale. During the years 1873 to 1876 at Yale University, American mathematical physicist Josiah Willard Gibbs, the first to be awarded a Ph.D. in engineering in the U.S., in a series of three papers, developed a mathematical-based, graphical methodology, for the study of chemical systems using the thermodynamics of Clausius. In 1882, German physicist Hermann von Helmholtz, published a founding thermodynamics paper, similar to Gibbs, but with more of an electro-chemical basis, in which he showed that measure of chemical affinity, i.e. the "force" of chemical reactions, is determined by the measure of the free energy of the reaction process. Following these early developments, the new science of chemical engineering began to develop.

Polymer properties are broadly divided into several classes based on the scale at which the property is defined as well as upon its physical basis. The most basic property of a polymer is the identity of its constituent monomers. A second set of properties, known as microstructure, essentially describe the arrangement of these monomers within the polymer at the scale of a single chain. These basic structural properties play a major role in determining bulk physical properties of the polymer, which describe how the polymer behaves as a continuous macroscopic material. Chemical properties, at the nano-scale, describe how the chains interact through various physical forces. At the macro-scale, they describe how the bulk polymer interacts with other chemicals and solvents. The identity of the monomer residues (repeat units) comprising a polymer is its first and most important attribute. Polymer nomenclature is generally based upon the type of monomer residues comprising the polymer. Polymers that contain only a single type of repeat unit are known as homopolymers, while polymers containing a mixture of repeat units are known as copolymers. Poly(styrene), for example, is composed only of styrene monomer residues, and is therefore classified as a homopolymer. Ethylene-vinyl acetate, on the other hand, contains more than one variety of repeat unit and is thus a copolymer. Some biological polymers are composed of a variety of different but structurally related monomer residues; for example, polynucleotides such as DNA are composed of a variety of nucleotide subunits.

The book examines the current economic and environmental consequences of expanding industrial activity including ozone depletion, acid rain, global warming and endocrine-disrupting chemicals. Emphasis has also on emerging chemical technologies based on green chemistry principles, water recycling and alternative fuels.

— Barry Griffin

Chapter 1

Water Chemistry

Water is continuously moving around the earth and constantly changing its form. It evaporates from land and water bodies and is also produced by all forms of life on Earth. This water vapour moves through the atmosphere, condenses to form clouds and precipitates as rain and snow. In time, the water returns to where it came from, and the process begins all over again. Although water is constantly moving, its total quantity on Earth's surface is constant.

Types of Water

Water is found in three different forms - liquid, solid or gas, depending on the temperature but it constantly changes from one form to another. Changes in temperature will determine which of these forms predominates in a particular area.

Liquid State

Water is usually encountered in the liquid state, because this is its natural state when temperatures are between 0° C and 100° C. 'Fresh' or drinking water is found as groundwater in underground aquifers, and on the surface in ponds, lakes, and rivers. Seas and oceans account for 97% of all water on Earth; but their waters contain dissolved salts and are therefore unfit to drink.

In regions of young volcanic activity, hot water emerges from the earth in hot springs (examples are Garampani in Assam and Badrinath in Uttaranchal). How does this phenomenon occur? Surface water percolates downward through the rocks below the Earth's surface to high-temperature regions surrounding a magma reservoir, either active, or recently solidified but still hot. There the water is heated, becomes less dense, and rises back to the surface through fissures and cracks.

Solid State

Ice is the frozen form of water. It occurs when temperatures are below 0°C (32°F). For a given mass, ice occupies 9% more volume than water, which is why when water enters cracks in rocks and freezes it causes the rocks to crack and split. Being less dense than water, ice floats. This property of ice is vital to aquatic life in cold regions. As the temperature drops, ice forms a protective, insulating layer on the surfaces of streams, pools and other water bodies, allowing water to remain liquid in the layers beneath and life to survive. Glaciers, icebergs, and ice caps are all frozen water.

Gas State

Water is found in the atmosphere in its gaseous form, water vapour. Steam is nothing but vapourized water. In certain hot water springs called geysers, jets of steam and hot water rise one hundred feet or more from the ground. Worldwide, the consumption of water is doubling every 20 years - more than twice the rate of increase in population.

A large amount of water is wasted in agriculture, industry, and urban areas. It has been estimated that with available technologies and better operational practices, agricultural water demand could be cut by about 50%, and that in urban areas by about 33% without affecting the quality or economics of life. But most governments do not have adequate laws or regulations to protect their water systems. Due to the increase in population there has been a rise in the demand for food, space for housing, consumer products, etc., which has in turn resulted in increased industrialization, urbanization, and demands in agriculture thereby leading to both river and groundwater contamination.

Treatment

Is a process of making water suitable for its application or returning its natural state. Thus, water treatment required before and after its application. The required treatment depends on the application. For example, treatment of greywater (from bath, dish and wash water) differs from the black water (from flush toilets). Composting toilet is not allowed in urban dwelling. Water treatment involves science, engineering, business, and art. The treatment may include mechanical, physical, biological, and chemical methods. As with any technology, science is the foundation, and engineering makes sure that the technology works as designed. The appearance and application of water is an art.

In terms of business, RGF Environmental, Water Energy Technologies, Aquasana Store, Vitech, Recalyx Industrial SDN BHD and PACE Chemicals ltd are some of many companies that offer various processes for water treatment. Millipore, a Fisher Scientific partner, offers many lines of products to produce *ultrapure water*, using a combination of active charcoal membranes, and reverse osmosis filter. Internet sites of these companies offer useful information regarding water.

An environmental scientist or consultant matches the service provider, modify if necessary, with the requirement.

- Natural Water includes some discussion on hard and soft water. Softening hard water for boiler, cooler, and domestic application is discussed therein. These treatments prepare water so that it is suitable for the applications.
- Water Biology deals with water and biology. Drinking water is part of making water suitable for living. Thus, this link gives some considerations to drinking water problems.
- There are many different industry types, and waters from various sources are usually treated before and after their applications. Pre-application treatment and wastewater treatment offer a special opportunity or challenge. Only a general consideration will be given to some industrial processes.
- General municipal and domestic wastewater treatment converts used water (waste) into environmentally acceptable water or even drinking water. Every urban centre requires such a facility.

General Water Treatment

All water treatments involve the removal of solids, bacteria, algae, plants, inorganic compounds, and organic compounds. Removal of solids is usually done by filtration and sediment. Bacteria digestion is an important process to remove harmful *pollutants*. Converting used water into environmentally acceptable water or even drinking water is wastewater treatment. Water in the Great Lakes Region is an organization dealing with the water resources. Ontario Clean Water Agency (OCWA) is a provincial Crown corporation in business to provide environmentally responsible and cost-efficient water and wastewater services. It currently operates more than 400 facilities for 200 municipalities.

Sewage Treatment

As a general discussion, let us look at a typical process in sewage treatment.

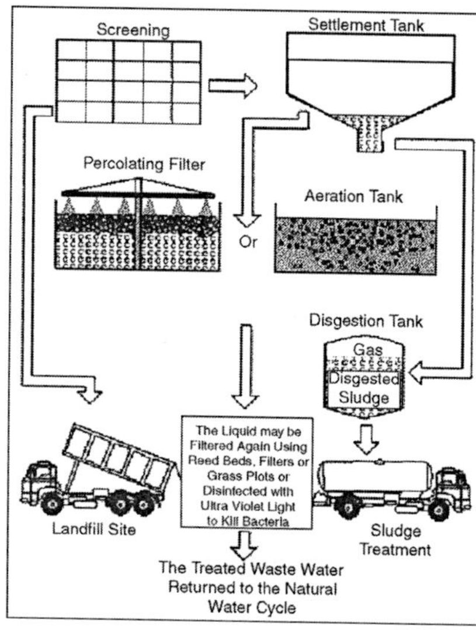

Sewage is screened to remove large solid chunks, which are disposed in landfill site. It flows over to the settlement tank to let the fine particles to settle. The settlement is called the activated sludge. The supernatant is then perclation filter and/or aerated. The water can be filtered again, and then disinfected (chlorinated in most cases). When there is no other complication, the water is returned to nature back to the ecological cycle. The sludge removed from the settlement is composed of living biological material. A portion of it may be returned to the aeration tank, but the raw sludge is digested by both microorganism. Anaerobic (without oxygen) and aerobic (with air) bacteria digestions are used. At the digestion stage, carbon dioxide, ammonia, and methane gases are evolved. Volume of the digested sludge is reduced, and it is acceptable as a fertilizer supplement in farming.

Wastewater Treatment

Although the sewage water may be discharged back to the ecological system after aerated digestion and perculationg filteration, but in some cases, further treatment is required. Some general consideration of water treatment is given below.

The criteria and standards for drinking water quality, organic compounds in waters, taste and order of water. Understandably, the standards change over the years. We have talked about drinking water in Water Biology.

Next deal with the removal of the following:

- Organics and inorganics by activated carbon
- Particulate matter by coagulation
- Particulate matter by filtration and sedimentation
- Hardness and other scale-forming substances
- Inorganic contaminants
- Corrosive substances
- Pathogenic (disease producing) bacteria, viruses, and protozoans (microorganisms).

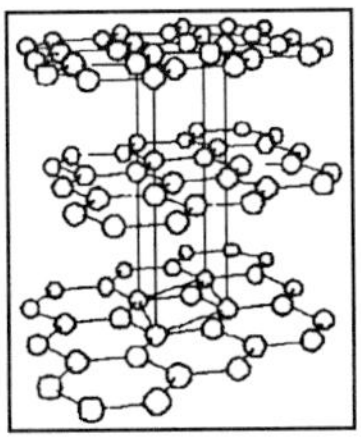

These items cover the chemistry, biology, and physics involved in the treatment of water. Some of these topics have been discussed in chemistry of water, physical properties of water, biology of water, and natural water..

Treatment by Carbon

Treatment by activated carbon is mostly due to adsorption or absorption. When a chemical species is adhered to the surface of a solid, it is an adsorption. When partial chemical bonds are formed between adsorbed species or when the absorbate got into the channels of the solids, we call it absorption. However, these two terms are often used to mean the same, because to distinguish one from type from the other is very difficult. Application of activated charcoal for the removal of undesirable order and taste in drinking water has been recognized at the dawn of civilization.

Using bone char and charred vegetation, gravel, and sand for the filtration of water for domestic application has been practised for thousands of years. Active research and production of activated charcoal was accelerated during the two world wars. The use of poison gas prompted the development of masks. They are still in use today. Charcoal absorbs many substances, ranging from colored organic particulates to inorganic metal ions. Charcoal has been used to remove the colour of raw sugar from various sources.

Charcoal consists of microcrystallines of graphite. The particles are so small in charcoal that they were considered amorphous. The crystal structure of graphite consists of layers of hexagonal networks, stacked on top of each other. Today, making activated carbon is a new and widely varied industry. Other molecules attach themselves to the porous surface and dangling carbons in these microcrystallines.

Carbon containing substances are charred at less than 900 K to produce carbon in the manufacture of activated carbon. However, the carbon is activated at 1200 K using oxidizing agent to selectively oxidize portions of the char to produce pores in the material. Because of the special process to produce used, these materials with high surface to mass ratio, they are called activated carbon rather than activated charcoal. Factors affecting the absorption are particle size, surface area, pore structure, acidity (pH), temperature, and the nature of the material to be absorbed. Usually, adsorption (absorption) equilibria and rate of adsorption must be considered for effective applications.

Coagulation, Flocculation and Sedimentation

Natural and wastewater containing small particulates. They are suspended in water forming a colloid. These particles carry the same charges, and repulsion prevents them from combining into larger particulates to settle. Thus, some chemical and physical techniques are applied to help them settle. The phenomenon is known as coagulation. A well known method is the addition of electrolyte.

Charged particulates combine with ions neutralizing the charges. The neutral particulates combine to form larger particles, and finally settle down. Another method is to use high-molecular-weight material to attract or trap the particulates and settle down together. Such a process is called flocculation. Starch and multiply charged ions are often used. Historically, dirty water is cleaned by treating with alum, $Al_2(SO_4)_3 \cdot 12\ H_2O$, and lime, $Ca(OH)_2$.

These electrolytes cause the pH of the water to change due to the following reactions:

$$Al_2(SO_4)_3 \cdot 12\ H_2O, \rightarrow Al_{(aq)}{}^{3+} + 3\ SO_{4(aq)}{}^{2-} + 12\ H_2O$$

$$SO_{4(aq)}{}^{2-} + H_2O \rightarrow HSO_{4(aq)}{}^{-} + OH^- \text{ (causing pH change)}$$

$$Ca(OH)_2 \rightarrow Ca_{(aq)}{}^{2+} + 2\ OH^- \text{(causing pH change)}$$

The slightly basic water causes $Al(OH)_3$, $Fe(OH)_3$ and $Fe(OH)_2$ to precipitate, bringing the small particulates with them and the water becomes clear. Some records have been found that Egyptians and Romans used these techniques as early as 2000 BC.

Suspension of iron oxide particulates and humic organic matter in water gives water the yellow muddy appearance. Both iron oxide particulates and organic matter can be removed from coagulation and flocculation. The description given here is oversimplified, and many more techniques have been applied in the treatment of water. Coagulation is a major application of lime in the treatment of wastewater. Other salts such as iron sulfates $Fe_2 (SO_4)_3$ and $FeSO_4$, chromium sulfate $Cr_2(SO_4)_3$, and some special polymers are also useful. Other ions such as sodium, chloride, calcium, magnesium, and potassium also affect the coagulation process. So do temperature, pH, and concentration.

Sedimentation let the water sit around to let the floculated or coagulated particles to settle out. It works best with relatively dense particles (*e.g.* silt and minerals), while flotation works better for lighter particles (*e.g.* algae, colour). A settling tank should be big enough so that it takes a long time (ideally 4 hours +) to get through. Inlets and outlets are designed so the water moves slowly in the tank. Long and narrow channels are installed to let the water to snake its way through the tank. The settled particles, sludge, must occasionally be removed from the tanks. The water is next ready to be filtered. Sedimentation is used in pre-treatment and wastewater treatment.

Filtration

Filtration is the process of removing solids from a fluid by passing it through a porous medium. Coarse, medium, and fine porous media have been used depending on the requirement. The filter media are artificial membranes, nets, sand filter, and high technological filter systems. The choice of filters depends on the required filtering speed and the *cleanness* requirement. The flow required for filtration can be achieved using gravity or pressure. In pressure filtration, one side of the filter medium is at higher pressure than that of the other so that the filter plane has a pressure drop. Some portion of this filter type must be enclosed in a container. The process of removing the clogged portion of the filter bed by reversing the flow through the bed and washing out the solid is called back washing. During this process, the solid must be removed out of the system, but otherwise the filters must be either replaced or taken out of service to be cleaned.

Aqua-Rain manufacture water filters as shown here. This unit consists of four filters. Regarding the filtering system, its techinical info gave the following statement. At the heart of the AquaRain Water Filtration System are Marathonr State-of-the-art ceramic elements utilizing a long-proven filtration process that is over 100 years old which will safely remove dangerous waterborne pathogens such as cysts (Cryptospordium, Giardia lamblia) and bacteria (E. coli, Samonelli typhus, etc.).

These innovative Marathonr ceramic elements are also filled with a high grade silvered granulated activated carbon (GAC). The GAC reduces pesticides, chemicals, chlorine, tastes and odours, while leaving the naturally occurring minerals found in the water unaffected. The units are designed for emergency and perhaps undeveloped countries.

Aeration

Bringing air into intimate contact with water for the purpose of exchanging certain components between the two phases is called aeration. Oxygenation is one of the purposes of aeration. Others are removal of volatile organic substances, hydrogen sulfide, ammonia, and volatile organic compounds.

A gas or substance dissolved in water may further react with water. Such a reaction is called hydration. Ionic substance dissolve due to hydration, for example:

$$HCl\ (g) + x\ H_2O = H(H_2O)_x{}^+ + Cl_{(aq)}{}^-$$

$$H_2S = H_{(aq)}{}^+ + HS_{(aq)}{}^-$$

These reactions are reversible, and aeration may also causes dehydration resulting in releasing the gas from water. Henry's law is applicable to this type of equilibrium for consideration.

Methods of aeration are:

- *Diffused aeration*: Air bubbles through water.
- *Spray aeration*: Water is sprayed through air.
- *Multiple-tray aeration:* Water flows through several trays to mix with air.
- *Cascade aeration*: Water flows downwards over many steps in the form of thin water falls.
- *Air stripping*: A combination of multiple tray and cascade technique plus random packed blocks causing water to mix thoroughly with air.

Reverse Osmosis Water Filter System

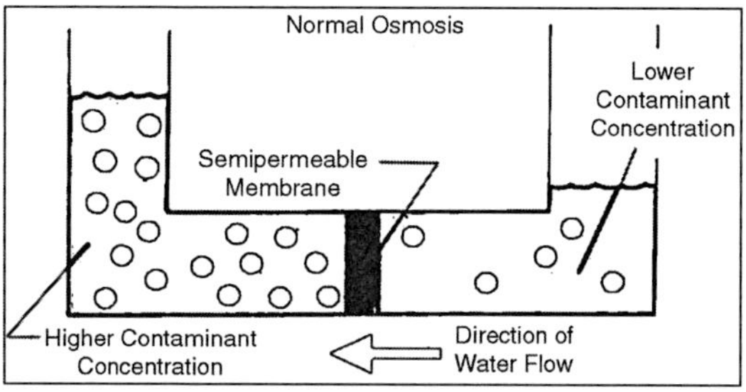

The dilute solution can be clean water whereas the concentrated solution contains undesirable solute (electrolyte or others). When a compartment containing a dilute solution is connected to another compartment containing a concentrated solution by a semi permeable membrane, water molecules move from the dilute solution to concentrated solution. This phenomenon is called osmosis. Pig bladders are natural semi permeable membranes.

As the water molecules migrate through the semi permeable membrane, water level in the solution will increase until the (osmotic) pressure prevents a net migration of water molecules in one direction. A pressure equivalent to the height difference is called the osmotic pressure. The illustration given on the right is from the PurePro, one of the many companies that manufacture reverse osmosis water filter devices.

Millipore also use this technique. By applying pressure in the higher concentration solution, water molecules migrate from a high concentration solution to a low concentration solution. This method is called reverse osmosis water filter system. The concept of reverse osmosis is illustrated in the diagram here from PurePro.In this technique, the membrane must be able to tolerate the high pressure, and prevent solute molecules to pass through. Regarding membranes, PurePro made the following statement:

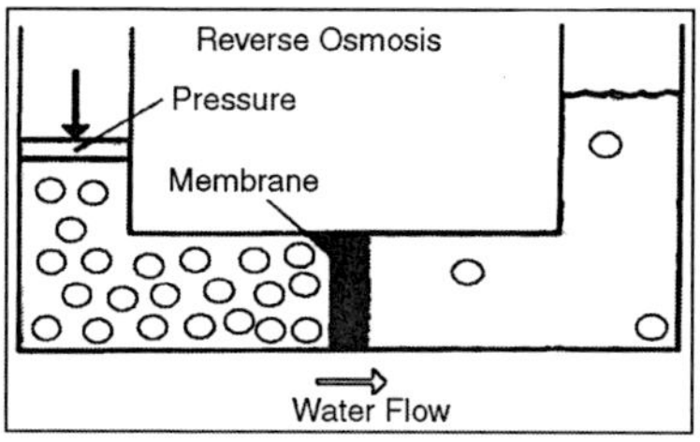

Semi permeable membranes have come a long way from the natural pig bladders used in the earlier osmosis experiments. Before the 1960's, these membranes were too inefficient, expensive, and unreliable for practical applications outside the laboratory.

Modern advances in synthetic materials have generally solved these problems, allowing membranes to become highly efficient at rejecting contaminants, and making them tough enough to withstand the greater pressures necessary for efficient operation.

This technology certainly works, and it has been used to convert salt (ocean or sea) water into fresh water. With this technique, the water with higher concentration is discharged. Thus, this technology is costly in regions where the water cost is high. Free Drinking Water also uses reverse osmosis filter system for domestic applications.

Uses of Water

Nearly everything we do in today's society requires water in some way. From making a car to preparing food in our homes, water is channeled for use by thousands of human industries.

A lot of the water today is transported via public supply. Public supply covers the water that is transported by water departments whether government owned, or private. These departments withdraw water from lakes, rivers, and reservoirs to wherever it needs to be, for domestic or industrial use, or any other use necessary. Around thirteen per cent of fresh water is affected by public supply.

- *The first of these is domestic use*: This covers water that is used in the home for private reasons, such as showering, toilets, cooking, washing and cleaning, and, when it happens, drinking. This can be either obtained from public supply or can be self supplied.
- *The second use is commercial use*: This is water for motels, hotels, restaurants, office buildings, and other commercial facilities, and institutions. This can also include fountain displays.
- *The third use for water and one of the most common (along with domestic use) is industrial use*: This includes water used in fabrication, processing, washing and cooling, and also can cover steel, chemical and allied products, paper and allied products, mining, and petroleum refining. This covers around five per cent of all water used. Industrial water use can sometimes use saline water, but 89% of the time its the same fresh water that we drink in our homes.

- *The fourth use is irrigation*: This is the when land is watered. This can cover the growing of crops, pastures, and maintaining vegetative growth in recreational lands, such as parks and golf courses.
- *The fifth use is thermoelectric power*: This is basically electrical power generation that it is not hydroelectric. Water is merely used in the process of creating this electricity.
- *The sixth use is mining*: This is not a huge use of water, however, often water is used to create a slurry something which allows the gold to be separated from the dirt.
- *The seventh use is livestock*: Livestock need water, and lots of it. One per cent of all fresh water used is used for livestock.
- *The eighth use is hydroelectric power*: This is power that is produced through falling water. The great thing about this however, is that the water does not have to be fresh. It does have to fall from a great hight at a high speed though. A great example of this is the snowy mountain scheme.

Water Waste

There is also a lot of water that is categorized as waste water:

- *Wastewater is used water*: It includes substances such as human waste, food scraps, oils, soaps and chemicals. In homes, this includes water from sinks, showers, bathtubs, toilets, washing machines and dishwashers. Businesses and industries also contribute their share of used water that must be cleaned.
- *Wastewater also includes storm run-off*: Although some people assume that the rain that runs down the street during a storm is fairly clean, it isn't. Harmful substances that wash off roads, parking lots, and rooftops can harm our rivers and lakes If it is not properly cleaned, water can carry disease. Since we live, work and play so close to water, harmful bacteria have to be removed to make water safe.

Agriculture mostly irrigation	69%
Industry	23%
Domestic use household,	
Drinking water and Sanitation	8%

These global averages vary a great deal between regions. In Countries, agriculture consumes 88 per cent of all water withdrawn for human use, while domestic use accounts for 7 per cent and industry for 5 per cent.

Agriculture

Almost 70 per cent of all available freshwater is used for agriculture. Pumping of groundwater by the world's farmers exceeds natural replenishment by at least 160 billion cubic metres a year. It takes an enormous amount of water to produce crops: one to three cubic metres to yield just one kilo of rice, and 1,000 tons of water to produce on ton of grain. Land in agricultural use has increased by 12% since the 1960s to about 1.5 billion hectares. Pasture and crops take up 37 per cent of the Earth's land area. Current global water withdrawals for irrigation are estimated at about 2,000 to 2,555 cubic kilometres per year. Poor drainage and irrigation practices have led to water logging and salinization of approximately 10 per cent of the world's irrigated lands (30 million hectares of the world's 255 million hectares of irrigated land) according to the UN Food and Agricultural

Organization (FAO)

A combination of Stalinization and water logging has degraded another 80 millions hectares. Agriculture is responsible for most of the depletion of groundwater, along with up to 70 per cent of the pollution. Both are accelerating. Many of the world's most important grain lands are consuming groundwater at unsustainable rates.

Industry

Water withdrawals for industry vary with the level of development:

- *World*: 22% of total water use.
- *High-income countries*: 59% of total water use
- *Low-income countries*: 8% of total water use

The annual water volume used by industry is estimated to rise from 752 km 3/year in 1995 to an estimated 1,170 km 3/year in 2025. In 2025, the industrial component is expected to represent about 24% of total freshwater withdrawal. Some 300-500 million tons of heavy metals, solvents, toxic sludge, and other wastes accumulate each year from industry. Industries based on organic raw materials are the most significant contributors to the organic pollutant load with the food sector being the most important polluter.

Contribution of the food sector to the production of organic water pollutant:

- *High income countries*: 40%
- *Low-income countries*: 54%

More than 80% of the world's hazardous waste is produced in the United States and other industrial countries. In developing countries, 70% of industrial wastes are dumped untreated into waters where they pollute the usable water supply.

Energy

World energy demand, especially for electricity, will increase greatly during this 21st century. Hydropower is the most important and widely-used renewable source of energy it represents 19% of total electricity production. Worldwide there are now about 45,000 large dams in operation. Canada is the largest producer of hydroelectricity, followed by the United States and Brazil.

Built to provide hydropower and irrigation water and to regulate river flow to prevent floods and draughts, they have had a disproportionate impact on the environment. Collectively, they have inundated more than 400,000 square kilometers of mostly productive land. Somewhere between 40 and 80 million people have been displaced by dams, forced to relocate to other, often less productive, land. A study by the World Commission on Dams, published in 2000, found that large dams have a very mixed record.

In 140 countries, dams provide cheap hydroelectric power. On a global scale, dams account for 19 per cent of the world's electricity generation and supply, through irrigation, almost 16 per cent of the world's food. Some dams continue to operate after 30-40 years, providing water and electricity. Hydropower plays a major role in reducing greenhouse gas emissions: developing ½ of the world's economically feasible hydropower potential could reduce greenhouse gases emissions by about 13%.

On the other hand large dams have led to the loss of forests and wildlife habitat and the loss of aquatic biodiversity - both upstream and downstream. Large dams have, in most cases, systematically failed to assess and account for the range of potential negative impacts on displaced and resettled communities.

With up to 80 million people displaced from their homes and many more living downstream suffering from unintended effects (eg. Loss of fisheries), mitigation efforts have, for the most part, been cosmetic and ineffective. According to the Commission, large dams may be on their way out: Mini-hydropower plants have proved to be far cheaper to build and more economical to run than originally forecast; plus they have minimal impacts on the environment. Better management to reduce the demand for water has great potential to reduce water stress and hydropower requirements.

Improved systems management, particularly for irrigated agriculture, has tremendous potential for reducing waste, while increasing the efficiency of irrigation systems. Demand for Water Increasing While freshwater supply is limited, demand risen as populations grow and consumption per capita increases. Global consumption of water is doubling every 20 years, more than twice the rate of human population growth, while pollution and over-extraction in many regions of the world has reduced the ability of supplies to meet demand.

During the last 70 years, the global population has tripled, but water withdrawals have increased over six times. Since 1940, annual global water withdrawals have increased by an average of nearly 3 per cent per year, while population growth has averaged between 1.5 and 2 per cent. According to the United Nations, more than one billion people on earth already lack access to fresh drink water. If current trends persists, by 2025 the demand for freshwater is expected to rise to 56 per cent more than the amount that is currently available.

More people mean increased water use and less available on a per capita basis. In 1989 there was some 9,000 cubic metres of freshwater per person available for human use. By 2000, that figures had dropped to 7,800 cubic metres and is expected to plummet to 5,100 cubic metres per person by 2025, when the global population is projected to reach 8 billion. The world's six billion people are already using about 54 per cent of all the accessible freshwater contained in rivers, lakes and underground aquifers. By 2025 the human's share will be 70 per cent, based on the population increase. If per capita consumption of water resources continues to rise at its current rate, humankind could be using over 90 per cent of all available freshwater within 25 years.

Water Shortages

A country experiences water stress when annual supplies drop below 1,700 cubic metres per person. When annual water supplies drop below 1,000 cubic metres per person, the country faces water scarcity for all or part of the year.

- In 1995, 31 countries containing 458 million people faced either water stress or scarcity.

- By 2025, according to projections made by Population Action International, more than 2.8 billion people in 48 countries will be facing water stress or scarcity.
- By 2050, the number of water short countries soars to 54, affecting 4 billion people, or 40 per cent of the projected global population.

Domestic Uses

Of course, some of the most important uses for water are at our homes. Domestic water use is water used for indoor and outdoor household purposes all the things you do at home: drinking, preparing food, bathing, washing clothes and dishes, brushing your teeth, watering the yard and garden, and even washing the dog. Water generally gets to our homes in one of two ways. Either it is delivered by a city/county water department (or maybe from a private company), or people supply their own water, normally from a well. Water delivered to homes is called "public supplied" and water that people supply themselves is called "self supplied." People who supply their own water almost always use ground. The vast majority of America's population gets their water delivered from a public-supply system.

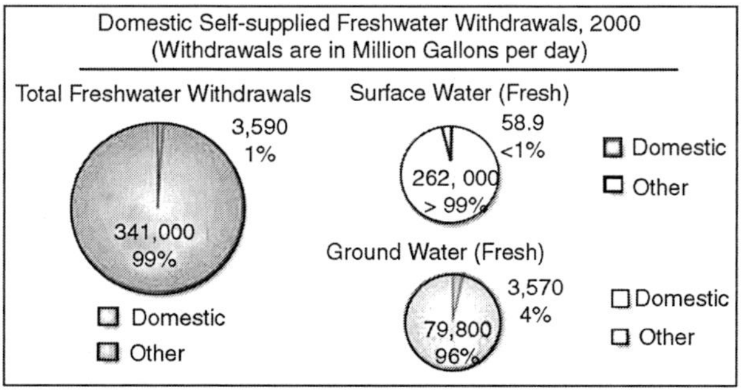

For 2000, withdrawals were an estimated 3,720 Mgal/d or 4,170 thousand acre-feet per year. Self-supplied domestic withdrawals were about 1 per cent of total freshwater withdrawals and about 2 per cent of total freshwater withdrawals for all categories excluding thermoelectric power. About 45 million people were self-supplied.

Ground water was the primary source of the water (98 per cent). Between 1995 and 2000, self-supplied domestic withdrawals increased about 10 per cent and the self-supplied domestic population increased almost 7 per cent. However, the self-supplied domestic population remained at 16 per cent of the total U.S. population

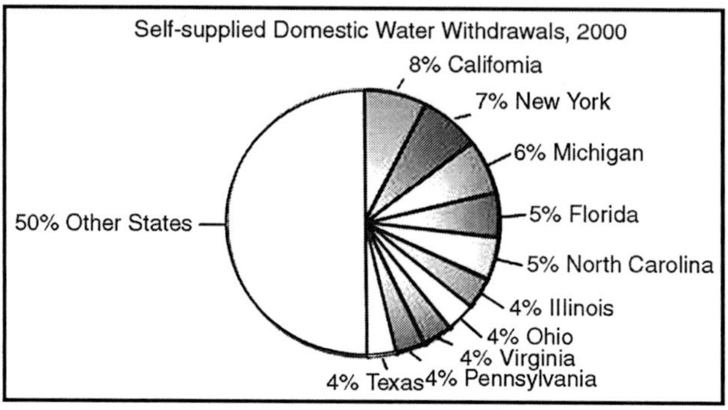

Trends in Domestic Water

In times past, when most people lived in rural areas, they had to find ways to supply their own water - often by drilling a well and pumping water to their homes. Not many city dwellers have a well in their backyards today. A public-water supply system, such as your local water department, nowadays delivers water to most homes.

The bar chart below shows the trend towards urbanization over the last 45 years. Notice how the blue bars (representing the millions of poeple served by a public water-supply system) keep going up while the green bars (representing the number of people who supply their own water) trended downward until 10 years ago. For the last 10 years, the percentage of the population relying on public-supply systems has remained at about 84 per cent.

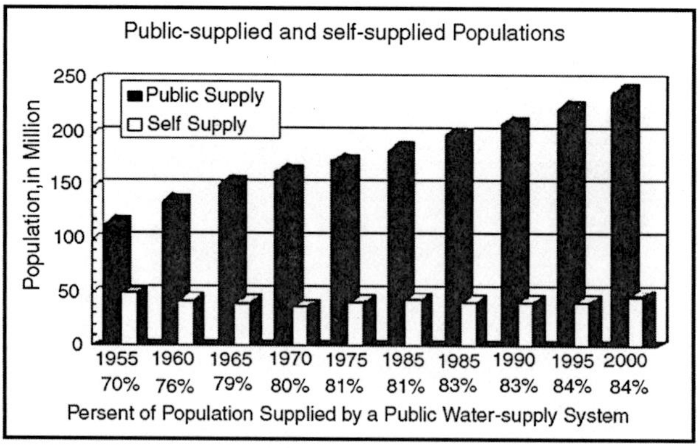

Irrigation Water for Drinking

Developing country's efforts to supply drinking water to their rural communities have focused primarily on digging deep tube wells and installing hand-pumps to exploit groundwater, which is free from harmful bacteria and parasites. But in large areas of South Asia, the Middle East and East Africa, groundwater is not an option because of high arsenic, fluoride, iron, or salt levels. Options for improving drinking water quality in areas where groundwater cannot be used have received very little attention. IWMI is trying to put this issue on the map by looking at the ways irrigation water can be made safer for human consumption.

IWMI is conducting a series of studies in Pakistan, Sri Lanka and Morocco that examine the links between irrigation management, domestic use of irrigation water and human health. This research has highlighted the need to bridge the gap between the irrigation and drinking water supply sectors. Currently, the potential for exploiting the health benefits of irrigation water is hindered by the lack of cooperation between drinking water providers and irrigation planners and managers. A more coordinated effort could improve water availability and quality. Especially if coupled with hygiene education, this could result in a significant improvement in community health.

Water Cycle

The water-use cycle is composed of the water cycle with the added influence of human activity. Dams, reservoirs, canals, aqueducts, withdrawal pipes in rivers, and groun-dwater wells all reveal that humans have a major impact on the water cycle. In the water-use cycle, water moves from a source to a point of use, and then to a point of disposition. The sources of water are either surface water or groundwater. Water is withdrawn and moved from a source to a point of use, such as an industry, restaurant, home, or farm. After water is used, it must be disposed of (or sometimes, reused). Used water is either directly returned to the environment or passes through a treatment processing plant before being returned.

Water Withdrawals

The Geological Survey compiles nationwide water-use data every 5 years. Even though discussions of water use typically focus on fresh-water use, saline water use also is important in countries. Some categories of water use, such as thermoelectric, industrial, and mining, use saline water, mainly for cooling generators in thermoelectric power plants. Industries and mines use saline water to cool machinery and to wash and transport products, mainly for cooling of machinery. After continual increases in the nation's total water withdrawals (fresh water and saline water) for the years reported from 1950 to 1980, withdrawals declined from 1980 to 1985 and remained fairly constant from 1985 to 1995. The 1995 estimate of total withdrawals (402,000 MGD) is about 2 per cent less than the 1990 estimate and nearly 10 per cent less than the peak year of 1980. Likewise, total fresh-water withdrawals for 1995 were about 8 per cent less than in 1980. The decline in withdrawals is especially significant in light of the fact that population posted an increase of 16 per cent during the same period.

Clearly as a nation, the countries are using its surface water and groundwater resources more efficiently. This decline signals that water use responds to economic and regulatory factors, and that the general public has an enhanced awareness of water resources and conservation issues.

The public supply, domestic, and livestock categories are the only categories to show continual increases from 1950 to 1995, largely because of continual increases in population. The increase of 4 per cent in public supply withdrawals from 1990 to 1995, compared to the population increase of 7 per cent served by public supply, indicates that conservation programmes have been effective in lowering public supply per capita use from about 184 gallons per day (GPD) in 1990 to 179 GPD in 1995.

More water continues to be withdrawn for thermoelectric power generation than for any other category. Withdrawals for thermoelectric power generation peaked in 1980 at 210,000 MGD and fluctuated around 190,000 MGD during 1985, 1990, and 1995.

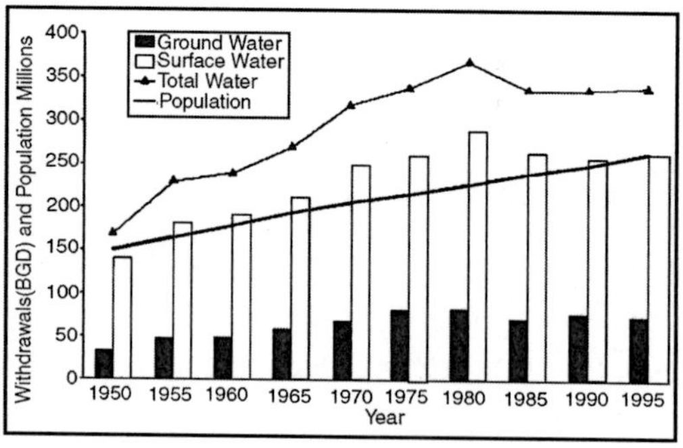

Figure: A Chart of Fresh-water

Industrial withdrawals declined from 1980 to 1995 after remaining about the same for the years reported between 1965 and 1980. Lower industrial withdrawals are the result of new industries and technologies that require less water, improved plant efficiencies, increased water recycling; changes in laws and regulations to reduce the discharge of pollutants; and conservation measures. Total irrigation withdrawals were about the same during the period from 1955 to 1960, then steadily increased from 1965 to 1980, and gradually decreased from 1980 to 1995.

Irrigation application rates vary from year to year and depend on annual rainfall, surface water availability, energy costs, farm commodity prices, application technologies, and conservation practices. Declining application rates are the result of more efficient irrigation systems and techniques. Also, application rates in the more humid eastern United States, where irrigated acreage continues to increase, tend to be lower than in the arid western.

Factors Affecting Water-Use Trends

Figure summarizes fresh-water withdrawals and population in the counry from 1950 to 1995. The general increase in water use from 1950 to 1980 and the decrease from 1980 to 1995 can be attributed, in part, to the following major factors:

- Most of the increases in water use from 1950 to 1980 were the result of expansion of irrigation systems and increases in energy development.
- The development of center-pivot irrigation systems and the availability of plentiful and inexpensive groundwater resources supported the expansion of irrigation systems.
- Higher energy prices in the 1970s, and large drawdown in groundwater levels in some areas increased the cost of irrigation water. In the 1980s, improved application techniques, increased competition for water, and a downturn in the farm economy reduced demands for irrigation water.
- The transition from water-supply management to water-demand management encouraged more efficient use of water.
- New technologies in the industrial sector that require less water, improved plant efficiencies, increased water recycling, higher energy prices, and changes in laws and regulations to reduce the discharge of pollutants resulted in decreased water use.
- The enhanced awareness by the general public of water resources and active conservation programmes in many states have contributed to reduced water demands.

Future Water Use

Water-use trends established over the past half century provide some basis for estimating future water demands. It seems likely that water withdrawals for public supply and domestic uses will continue to increase as population increases. Higher water prices and active water conservation progra-mmes, however, may reduce the per capita use rates. With increased competition for water for in stream uses, such as river-based recreation, aesthetic enjoyment, fish and wildlife habitat, and hydroelectric power, along with higher municipal uses, irrigators will have increasing difficulty competing economically for available water supplies.

In the United States, many existing sources of water are being stressed by withdrawals from aquifers and diversions from rivers and reservoirs to meet the needs of homes, cities, farms, and industries. Consideration must also be given to leaving water in the streams and rivers to meet environmental, fish and wildlife, and recreational needs.

Method of Water Softening

Water Hardness

Carbon dioxide reacts with water to form carbonic acid which at ordinary environmental pH exists mostly as bicarbonate ion. Microscopic marine organisms take this up as carbonate to form calcite skeletons which, over millions of years, have built up extensive limestone deposits. Groundwater, made slightly acidic by CO_2 (both that absorbed from the air and from the respiration of soil bacteria) dissolve the limestone, thereby acquiring calcium and bicarbonate ions and becoming "hard".

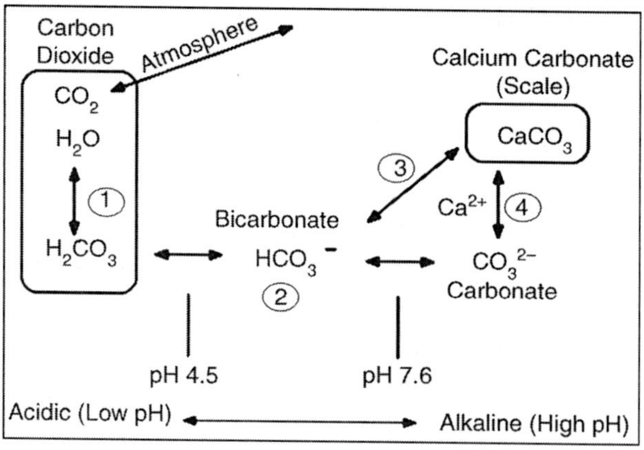

If the HCO_3^- concentration is sufficiently great, the combination of processes calcium carbonate ("lime scale") to precipitate out on surfaces such as the insides of pipes. (Calcium bicarbonate itself does not form a solid, but always precipitates as $CaCO_3$.)

These "hardness ions" cause two major kinds of problems. First, the metal captions react with soaps, causing them to form an unsightly precipitate— the familiar "bathtub ring". More seriously, the calcium and magnesium carbonates tend to precipitate out as adherent solids on the surfaces of pipes and especially on the hot heat exchanger surfaces of boilers. The resulting scale buildup can impede water flow in pipes. In boilers, the deposits act as thermal insulation that impedes the flow of heat into the water; this not only reduces heating efficiency, but allows the metal to overheat, which in pressurized systems can lead to catastrophic failure.

Types of Water Hardness

Temporary Hardness

This refers to hardness whose effects can be removed by boiling the water in an open container. Such waters have usually percolated though limestone formations and contain bicarbonate HCO_3^- along with small amounts of carbonate CO_3^{2-} as the principal negative ions. Boiling the water promotes the reaction,

$$2\ HCO_3^{-}\ '\rightarrow\ CO_3^{2-} + CO_2$$

by driving off the carbon dioxide gas. The CO_3^{2-} reacts with Ca^{2+} or Mg^{2+} ions, to form insoluble calcium and magnesium carbonates which precipitate out. By tying up the metal ions in this way, the amounts available to form soap scum are greatly reduced.

Permanent Hardness

Waters than contain other anions such as chloride or sulfate cannot be remediate by boiling, and are said to be "permanently" hard. The only practical treatment is to remove all the ions, normally by the method described below.

Conventional Water Softening

Most conventional water-softening devices depend on a process known as *ion-exchange* in which "hardness" ions trade places with sodium and chloride ions that are loosely bound to an *ion-exchange resin* or a *zeolite* (many zeolite minerals occur in nature, but specialized ones are often made artificially.)

The illustration in fig depicts a negatively-charged zeolite to which [positive] sodium ions are attached. Calcium or magnesium ions in the water displace sodium ions, which are released into the water. In a similar way, positively-charged zeolites bind negatively-charged chloride ions (Cl^-), which get displaced by bicarbonate ions in the water. As the zeolites become converted to their Ca^{2+} and HCO_3^- forms they gradually lose their effectiveness and must be regenerated.

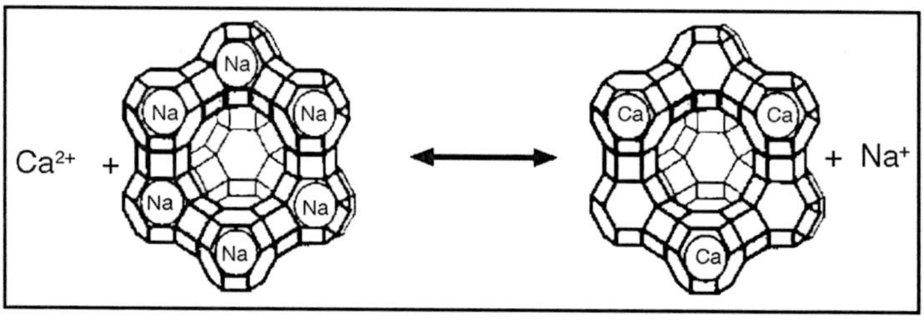

This is accomplished by passing a concentrated brine solution though them, causing the above reaction to be reversed. Herein lies one of the drawbacks of this process: most of the salt employed in the regeneration process gets flushed out of the system and is usually released into the soil or drainage system something that can have damaging consequences to the environment, especially in arid regions. For this reason, many jurisdictions prohibit such release, and require users to dispose of the spent brine at an approved site or to use a commercial service company.

Alternative Methods

The great economic importance of water softening has created a large and thriving industry that utilizes a number of proven methods based on well established scientific principles. It has also unfortunately attracted a variety of operators offering technologies that are purported to be better, less expensive, easier to install, or "chemical-free", but which have never been validated scientifically and whose principles of operation are largely unexplained by the known laws of chemistry. This does not mean that such schemes cannot work (after all, we can use theory to show that under idealized conditions, water can never boil and it can never rain!), but it should inspire a good degree of skepticism. Most of the statements supporting alternative water treatment methods come from those who have a commercial interest in these devices, they are not supported by credible and independently verifiable performance data, and the explanations they offer for how they work reveal such a weak understanding of basic chemistry on the part of their authors that it is difficult to have much confidence in them.

Against this, there is some anecdotal evidence that certain magnetic and electromagnetic devices can be effective in preventing scale formation in hard water systems. It is very difficult to judge such claims, which are almost never based on tests that are well enough described to allow others to evaluate them and to verify the results. While the lack of "scientific" evidence does not in itself invalidate a claim for the efficacy of a device, it should make one hesitate to accept it without some guarantee of performance.

Distillation

Distillation is probably the oldest method of water purification. Water is first heated to boiling. The water vapour rises to a condenser where cooling water lowers the temperature so the vapour is condensed, collected and stored. Most contaminants remain behind in the liquid phase vessel. However, there can sometimes be what is called carry-overs in the water that is distilled.

Organics such as herbicides and pesticides, with boiling points lower than 100 Â°C cannot be removed efficiently and can actually become concentrated in the product water. Another disadvantage is cost. Distillation requires large amounts of energy and water. Distilled water can also be very acidic, having a low pH, thus should be contained in glass. Since there is not much left, distilled water is often called "hungry" water. It lacks oxygen and minerals and has a flat taste, which is why it is mostly used in industrial processes.

Table: Distillation

Advantages	*Disadvantages*
Removes a broad range of contaminants Reusable ensure purity	Some contaminants can be carried into the condensate Requires careful maintenance to Consumes large amounts of energy

Ion Exchange

The ion exchange process percolates water through bead-like spherical resin materials (ion-exchange resins). Ions in the water are exchanged for other ions fixed to the beads. The two most common ion-exchange methods are softening and deionization. Softening is used primarily as a pretreatment method to reduce water hardness prior to reverse osmosis (RO) processing. The softeners contain beads that exchange two sodium ions for every calcium or magnesium ion removed from the "softened" water.

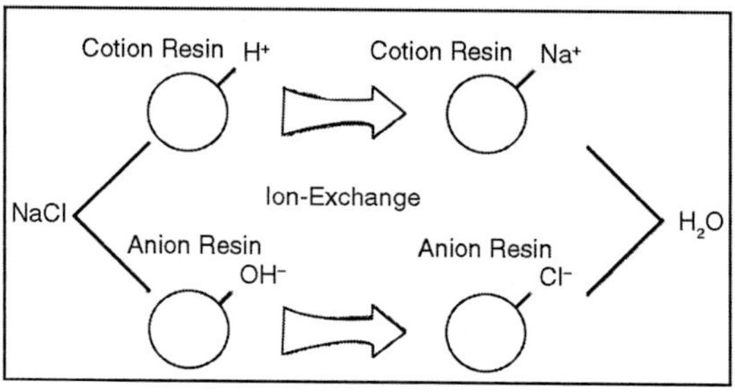

Deionization (DI) beads exchange either hydrogen ions for cations or hydroxyl ions for anions. The cation exchange resins, made of styrene and divinylbenzene containing sulfonic acid groups, will exchange a hydrogen ion for any cations they encounter (*e.g.*, Na+, Ca++, Al+++).

Similarly, the anion exchange resins, made of styrene and containing quaternary ammonium groups, will exchange a hydroxyl ion for any anions (*e.g.*, Cl-). The hydrogen ion from the cation exchanger unites with the hydroxyl ion of the anion exchanger to form pure water.

These resins may be packaged in separate bed exchangers with separate units for the cation and anion exchange beds. Or, they may be packed in mixed bed exchangers containing a mixture of both types of resins. In either case, the resin must be "regenerated" once it has exchanged all its hydrogen and/or hydroxyl ions for charged contaminants in the water. This regeneration reverses the purification process, replacing the contaminants bound to the DI resins with hydrogen and hydroxyl ions.

Deionization can be an important component of a total water purification system when used in combination with other methods discussed in this primer such as RO, filtration and carbon adsorption. DI systems effectively remove ions, but they do not effectively remove most organics or microorganisms.

Microorganisms can attach to the resins, providing a culture media for rapid bacterial growth and subsequent pyrogen generation. The advantages and disadvantages of this technology are summarized below.

Table: Deionization

Advantages	*Disadvantages*
Removes dissolved inorganics effectively.	Does not effectively remove particles, pyrogens or bacteria.
Regenerable (service deionization).	DI beds can generate resin particles and culture bacteria.
Relatively inexpensive initial capital investment.	High operating costs over long-term.

Carbon Adsorption

Carbon absorption is a widely used method of home water treatment because of its ability to improve water by removing disagreeable tastes and odours, including objectionable chlorine. Activated carbon effectively removes many chemicals and gases, and in some cases it can be effective against microorganisms. However, generally it will not affect total dissolved solids, hardness, or heavy metals. Only a few carbon filter systems have been certified for the removal of lead, asbestos, cysts, and coli form.

Activated carbon is created from a variety of carbon-based materials in a high-temperature process that creates a matrix of millions of microscopic pores and crevices. One pound of activated carbon provides anywhere from 60 to 150 acres of surface area. The pores trap microscopic particles and large organic molecules, while the activated surface areas cling to, or adsorb, small organic molecules.

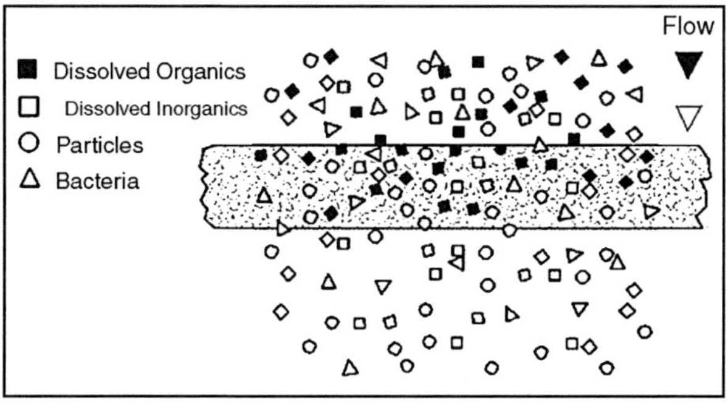

The ability of an activated carbon filter to remove certain microorganisms and certain organic chemicals, especially pesticides, THMs (the chlorine by-product), trichloroethylene (TCE), and PCBs, depends upon several factors, such as the type of carbon and the amount used, the design of the filter and the rate of water flow, how long the filter has been in use, and the types of impurities the filter has previously removed.

The carbon adsorption process is controlled by the diameter of the pores in the carbon filter and by the diffusion rate of organic molecules through the pores. The rate of adsorption is a function of the molecular weight and the molecular size of the organics. Certain granular carbons effectively remove chloramines. Carbon also removes free chlorine and protects other purification media in the system that may be sensitive to an oxidant such as chlorine.

Microporous Basic Filtration

There are three types of micro porous filtration: depth, screen and surface. Depth filters are matted fibers or materials compressed to form a matrix that retains particles by random adsorption or entrapment. Screen filters are inherently uniform structures which, like a sieve, retain all particles larger than the precisely controlled pore size on their surface. Surface filters are made from multiple layers of media. When fluid passes through the filter, particles larger than the spaces within the filter matrix are retained, accumulating primarily on the surface of the filter.

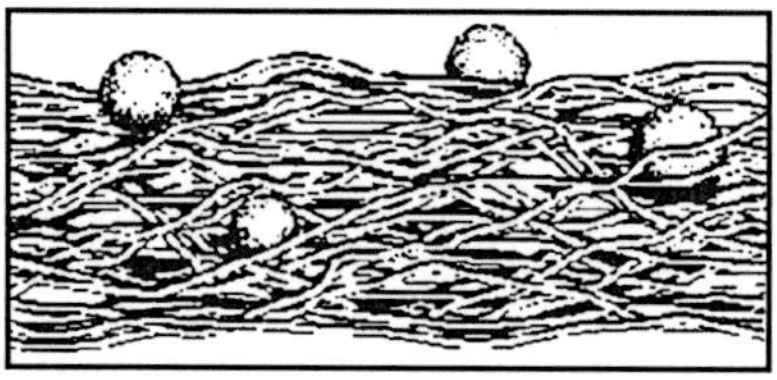

The distinction between filters is important because the three serve very different functions. Depth filters are usually used as prefilters because they are an economical way to remove 98% of suspended solids and protect elements downstream from fouling or clogging. Surface filters remove 99.99% of suspended solids and may be used as either prefilters or clarifying filters. Micro porous membrane (screen) filters are placed at the last possible point in a system to remove the last remaining traces of resin fragments, carbon fines, colloidal particles and microorga-nisms.

Ultra Filtration

A micro porous membrane filter removes particles according to pore size. By contrast, an ultra filtration (UF) membrane functions as a molecular sieve. It separates dissolved molecules on the basis of size by passing a solution through an infinitesimally fine filter.

The ultra filter is a tough, thin, selectively permeable membrane that retains most macromolecules above a certain size including colloids, microorganisms and pyroxenes. Smaller molecules, such as solvents and ionized contaminants, are allowed to pass into the filtrate. Thus, UF provides a retained fraction (retentate) that is rich in large molecules and a filtrate that contains few, if any, of these molecules.

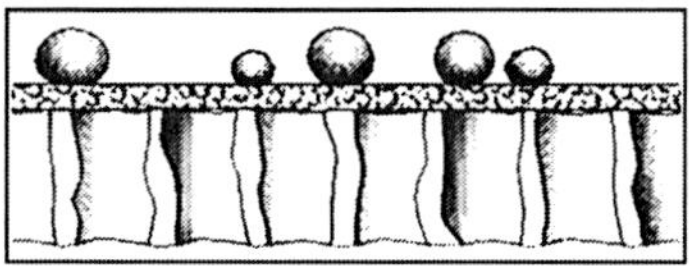

Ultra filters are available in several selective ranges. In all cases, the membranes will retain most, but not necessarily all, molecules above their rated size.

Reverse Osmosis

Reverse osmosis (RO) is the most economical method of removing 90% to 99% of all contaminants. The pore structure of RO membranes is much tighter than UF membranes. RO membranes are capable of rejecting practically all particles, bacteria and organics >300 daltons molecular weight (including pyrogens). In fact, reverse osmosis technology is used by most leading water bottling plants. Natural osmosis occurs when solutions with two different concentrations are separated by a semi-permeable membrane. Osmotic pressure drives water through the membrane; the water dilutes the more concentrated solution; and the end result is equilibrium. In water purification systems, hydraulic pressure is applied to the concentrated solution to counteract the osmotic pressure. Pure water is driven from the concentrated solution and collected downstream of the membrane. Because RO membranes are very restrictive, they yield slow flow rates. Storage tanks are required to produce an adequate volume in a reasonable amount of time.

RO also involves an ionic exclusion process. Only solvent is allowed to pass through the semi-permeable RO membrane, while virtually all ions and dissolved molecules are retained (including salts and sugars). The semi-permeable membrane rejects salts (ions) by a charge phenomena action: the greater the charge, the greater the rejection. Therefore, the membrane rejects nearly all (>99%) strongly ionized polyvalent ions but only 95% of the weakly ionized monovalent ions like sodium. Reverse osmosis is highly effective in removing several impurities from water such as total dissolved solids (TDS), turbidity, asbestos, lead and other toxic heavy metals, radium, and many dissolved organics. The process will also remove chlorinated pesticides and most heavier-weight VOCs. Reverse osmosis and activated carbon filtration are complementary processes. RO is the most economical and efficient method for purifying tap water if the system is properly designed for the feed water conditions and the intended use of the product water. RO is also the optimum pretreatment for reagent-grade water polishing systems.

In addition, Reverse osmosis treatment is an insurance policy against nuclear radiation such as radioactive plutonium or strontium in the drinking water. If one lives near a nuclear power plant, this is a key way to ensure the household is drinking the best water for their health.

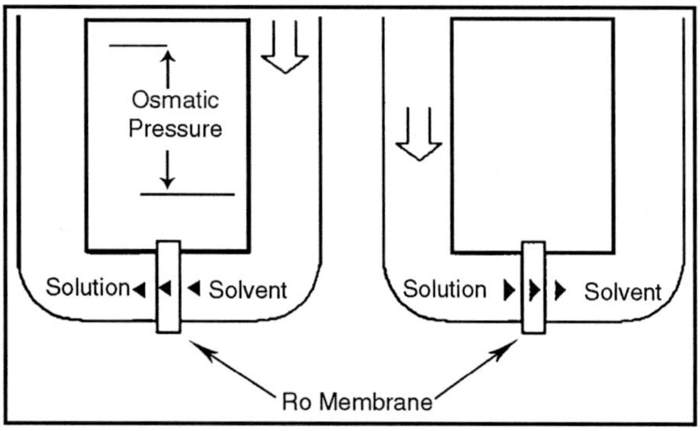

Ultraviolet (UV) Radiation

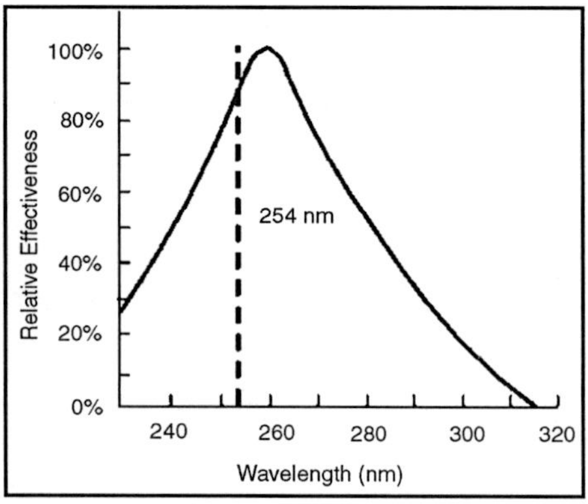

Ultraviolet radiation has widely been used as a germicidal treatment for water. Mercury low pressure lamps generating 254 nm UV light are an effective means of sanitizing water. The adsorption of UV light by the DNA and proteins in the microbial cell results in the inactivation of the microorganism.

Recent advances in UV lamp technology have resulted in the production of special lamps which generate both 185 nm and 254 nm UV light. This combination of wavelengths is necessary for the photo oxidation of organic compounds. With these special lamps, Total Organic Carbon (TOC) levels in high purity water can be reduced to 5 ppb.

Water Purification Systems

Because each purification technology removes a specific type of contaminant, none can be relied upon to remove all contaminants to the levels required for critical applications. A well-designed water purification system uses a combination of purification technologies to achieve final water quality.

Each of the purification technologies must be used in an appropriate sequence to optimize their particular removal capabilities. The schematic below shows a central laboratory water purification system designed to produce water for critical applications. The first step is pretreatment equipment specifically designed to remove contaminants in the feed water. Pretreatment removes contaminants that may affect purification equipment located downstream especially reverse osmosis (RO) systems. Examples of pretreatment are: carbon filters (or tanks) for chlorine removal, particulate filters for sediment/ silt/particulate removal, and softening agents to remove minerals that cause “hard” water.

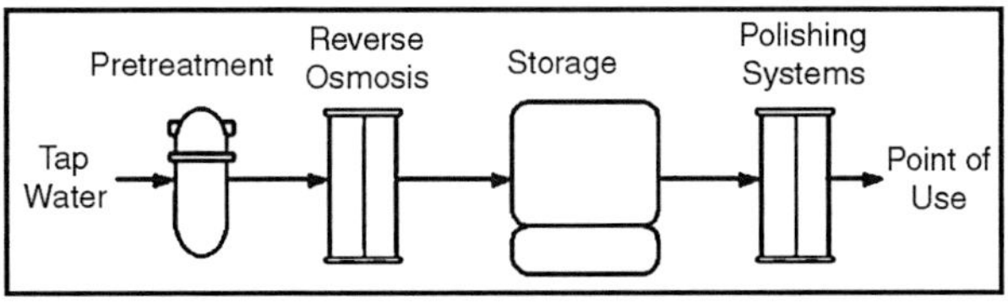

The next purification step is RO: RO removes 90 to 99% of all the contaminants found in water. It is the heart of any well designed water purification system because it effectively removes a broad range of contaminants. However, the tight porosity of the RO membrane limits its flow rate. Therefore, a storage container is used to collect water from the system and distribute it to other points-of-use such as polishing systems.

Polishing systems purify pretreated water, such as RO water, by removing trace levels of any residual contaminants. Polishing elevates the quality of pretreated water to “Type I” or “ultra pure” water. A polishing system is designed to remove residual traces of impurities from water already pretreated by some other means (such as reverse osmosis or deionization). Treating raw tap water using such a system would quickly exhaust its capacity and affect final quality.

A typical polishing system may consist of activated carbon, mixed-bed deionization, organic scavenging mixtures and 0.22 Âµm final filtration. Systems can also be enhanced with ultra filtration, ultraviolet oxidation or other features for use in specific applications.

This combination of purification technologies, combined with proper pretreatment, will produce water that is virtually free of ionic, organic and microbial contamination.

Water Softening Device and Method

Field of the Invention

The present invention relates to the field of fabric cleaning methods. The invention is concerned with a water softening device for application in automatic washing machines, more particularly, a water softening device based on capacitive deionization in a flow-through capacitor for obtaining water that is suitable for use with detergent products having low environmental impact.

Background of the Invention

In recent years one has become increasingly aware of the impact of human activities on the environment and the negative consequences this may have. Ways to reduce, reuse and recycle resources are becoming more important. Fabric cleaning is one of the many household activities with a significant environmental impact. This is partly caused by the use of conventional detergent products, which tend to be relatively complex compositions with a variety of ingredients. Over the years some ingredients have been banned by legislation in certain countries because of their adverse environmental effects. Examples include certain nonionic surfactants and builders such as phosphates. The use of phosphates in detergents has been linked to increased levels of phosphates in surface waters. The resulting eutrophication is thought to cause an increased growth of algae. The increased algae growth in stagnant surface water leads to oxygen depletion in lower water layers, which in turn causes general reduction of overall water quality.

Although some ingredients in conventional laundry detergent products may have a limited environmental effect, the energy involved in the production thereof influences the environmental impact during their life cycle negatively. Life cycle analysis typically estimates the environmental impact of a product during the different phases such as production of raw material, production of the product itself, distribution to the end user, use of the product by for example the consumer and the disposal after use. Environmental impact may include factors like eutrophication, green house effect, acidification and photo-chemical oxidant formation. With respect to laundry detergent products, extra ingredients necessarily add cost, volume and weight to the product, which in turn requires more packaging material and transport costs. Extra ingredients usually require a more complex production process.

However, it is difficult to reduce the number and/or amount of the ingredients without reducing the cleaning efficiency. One of the most bulky ingredients of common laundry detergents are so-called builders like for example zeolites, phosphates, soaps and carbonates. Builders are added to laundry detergent formulations for their ability to sequester hardness-ions like Ca^{2+} and Mg^{2+}. The reduction of hardness ions is required in order to prevent the deposition of calcium soaps in the soil, to prevent the precipitation of anionic surfactants, to maximise colloid stability and to reduce the calcium-soil-substrate-interaction and soil-soil interaction and hence to improve soil removal. Apart from their positive effects, common builders also may have negative effects on laundry cleaning processes. Builders often generate insoluble materials in the wash either as such or by formation of precipitates. For example, zeolites are insoluble and may cause incrustation of fabrics and precipitates of calcium-builder-complex result in higher redepositioning.

In short, builders are required for sequestering hardness ions to improve wash efficiency, but have a negative environmental effect and generate insoluble precipitates that may cause redepositioning on the fabric articles and thereby reduce the wash efficiency. However, the requirement for builder material may be reduced when soft water is used in the washing process.

Different methods are known in the art to produce soft water by sequestering hardness-ions like Ca^{2+} and Mg^{2+} from tap water, for instance by ion-exchange. In WO01/30229, a system is described, which utilizes a built-in ion-exchange system to remove calcium and magnesium ions from the water supply. However, the ion-exchange material requires regular regeneration. For application in a common type of automatic washing machine, vast amounts of *e.g.* salt solution would be required for the regeneration of the ion-exchanger, thereby undoing the effect of the reduction of builder chemicals in the detergent. Further disadvantages of ion-exchange are the limited life-time of the ion-exchange resin and/or the required volume of resin for the production of the amount of soft water required in a washing machine.

Another water softening method is electronic deionization (EDI), which combines ion exchange and electro dialysis, as described in co-pending application 04076353.4. Although this method does not require regeneration chemicals, the other disadvantages of the ion-exchange resin remain as indicated above. Furthermore, EDI is a complicated technology, that is difficult to operate in a robust manner over a long time period, as required in house-hold appliances. A known method for water treatment is capacitive deionization, using a flow through capacitor (FTC) as among others described in U.S. Said method comprises the use of an electrically regenerable electrochemical cell for capacitive deionization and electrochemical purification and regeneration of the electrodes including two end plates, one at each end of the cell. By polarizing the cell, ions are removed from the electrolyte and are held in the electric double layers at the electrodes. The cell can be (partially) regenerated electrically to desorbs such previously removed ions.

The regeneration could be carried out without added chemical substances. The term charge barrier refers to a layer of material which is permeable or semi-permeable and is capable of holding an electric charge. Pore volume ions are retained, or trapped, on the side of the charge barrier towards which the like-charged ions migrate. Generally, a charge barrier functions by forming a concentrated layer of ions. The effect of forming a concentrated layer of ions balances out, or compensates for, the losses ordinarily associated with pore volume ions. This effect allows a large increase in ionic efficiency, which in turn allows energy efficient purification of concentrated fluids. Using the charge barrier flow-through capacitor in the purification of water has been observed at an energy level of less than 1 joules per coulomb ions purified, for example, 0.5 joules per coulomb ions purified, with an ionic efficiency of over 90%.

It is an object of the present invention to find a cost-effective method having low environmental impact for removing hardness ions from tap water. It is another object of the invention to find a cost-effective method having low environmental impact both for removing hardness ions from tap water and for modifying the pH. Another object of the present invention is to find a method for removing hardness ions from tap water and for modifying the pH of said water in a manner that is robust, long lasting, and convenient and user friendly to consumers. It is a further object of the invention to find a method to remove hardness ions from the tap water, without the need for added chemicals or vast amounts of water. It is another object of the invention to find a method to remove hardness ions from a softening device, without the need for added chemicals or vast amounts of water. Yet another object of the invention is to find a suitable method for treating tap water such that water is obtained that is suitable for use with a low environmental impact detergent product (LEIP, as defined herein), in fabric cleaning methods. A still further object of the invention is to find a cleaning method wherein water obtained from such a water treatment method can be suitably used with a LEIP in in-home cleaning appliances, such as a fabric washing machine.

Present Invention

Accordingly, the present invention provides a water softening device for application in a household appliance comprising a flow-through capacitor for the production of wash amplified water (WAW) from tap water, said WAW having less than 50 FH, and being suitable for use in said appliance when the device is in operation; whereby the configuration of the device is such that the capacitor can be regenerated, whereby no added substances are used; and a pH modifier that can be fed with tap water or softened water, and is able to split this water in an alkaline and an acidic water stream; and wherein the ratio between WAW and waste water from the flow-through capacitor is from 5:1 to 100:1.

The invention also provides a laundering process for the cleaning of fabric articles wherein water softening device 30 according to the invention is used. The invention further provides a water softening process wherein the device of the invention is used and wherein the anions present in the feed water are attracted to the anode plates and cations in the water are attracted to the cathode plates when the device is in operation. For the purpose of the present invention, the feed water is defined to be water having a conductivity of more than 50 micro Siemens cm^{H1}, preferably more than 100 micro Siemens cm^{H1} and more preferably more than 200 micro Siemens cm^{H1}. For practical reasons, the feed water is desirably tap water from the main, having a water hardness of at least 7° FH. Preferably, the cleaning method of the invention is carried out in a fabric or dish washing machine, more preferably a fabric washing machine. In view of this, it is desirable that the wash amplified water has a pH of above 8.5, more preferably above 9.5. The cleaning method of the invention is particularly suitable for in-home use and the wash amplified water obtained from said method is suitable for use in a household-cleaning appliance.

These and other aspects, features and advantages will become apparent to those of ordinary skill in the art from a reading of the following detailed description and the appended claims. For the avoidance of doubt, any feature of one aspect of the present invention may be utilized in any other aspect of the invention. It is noted that the examples given in the description below are intended to clarify the invention and are not intended to limit the invention to those examples per se. Similarly, all percentages are weight/weight percentages of the low environmental detergent product composition unless otherwise indicated. Numerical ranges expressed in the format "from x to y" are understood to include x and y. When for a specific feature multiple preferred ranges are described in the format "from x to y", it is understood that all ranges combining the different endpoints are also contemplated.

Boiler Feed Water

A boiler is a device for generating steam, which consists of two principal parts: the furnace, which provides heat, usually by burning a fuel, and the boiler proper, a device in which the heat changes water into steam. The steam or hot fluid is then recirculated out of the boiler for use in various processes in heating applications.

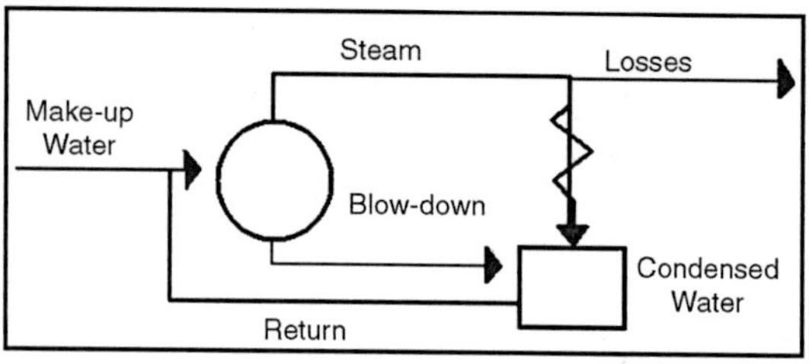

The boiler receives the feed water, which consists of varying proportion of recovered condensed water (return water) and fresh water, which has been purified in varying degrees (make up water). The make-up water is usually natural water either in its raw state, or treated by some process before use. Feed-water composition therefore depends on the quality of the make-up water and the amount of condensate returned to the boiler.

The steam, which escapes from the boiler, frequently contains liquid droplets and gases. The water remaining in liquid form at the bottom of the boiler picks up all the foreign matter from the water that was converted to steam. The impurities must be blown down by the discharge of some of the water from the boiler to the drains. The permissible percentage of blown down at a plant is strictly limited by running costs and initial outlay.

Scalimng

Boiler scale is caused by impurities being precipitated out of the water directly on heat transfer surfaces or by suspended matter in water settling out on the metal and becoming hard and adherent. Evaporation in a boiler causes impurities to concentrate. This interferes with heat transfers and may cause hot spots. Leading to local overheating. Scaling mechanism is the exceeding of the solubility limits of mineral substances due to elevated temperature and solids concentration at the tube/water interface. The deposition of crystalline precipitates on the walls of the boiler interferes with heat transfer and may cause hot spots, leading to local overheating. The less heat they conduct, the more dangerous they are.

Common feed water contaminants that can form boiler deposits include calcium, magnesium, iron,aluminum, and silica. Scale is formed by salts that have limited solubility but are not totally insoluble in boiler water. These salts reach the deposit site in a soluble form and precipitate.

The values corresponding to their thermal conductivity are:

Steel	15	kcal/m^2.h per degree C
$CaSO_4$	1-2	kcal/m^2.h per degree C
$CaCO_3$	0.5-1	kcal/m^2.h per degree C
SiO_2	0.2-0.5	kcal/m^2.h per degree C

Scaling is mainly due to the presence of calcium and magnesium salts (carbonates or sulphates), which are less soluble hot than cold, or to the presence of too high concentration of silica in relation to the alkalinity of the water in the boiler.

A carbonate deposit is usually granular and sometimes of a very porous nature. The crystals of calcium carbonate are large but usually are matted together with finely divided particles of other materials so that the scale looks dense and uniform. Dropping it in a solution of acid can easily identify a carbonate deposit. Bubbles of carbon dioxide will effervesce from the scale. A sulphate deposit is much harder and more dense than a carbonate deposit because the crystals are smaller and cement together tighter. A Sulphate deposit is brittle, does not pulverize easily, and does not effervesce when dropped into acid. A high silica deposit is very hard, resembling porcelain.

The crystal of silica is extremely small, forming a very dense and impervious scale. This scale is extremely brittle and very difficult to pulverize. It is not soluble in hydrochloric acid and is usually very light coloured. Iron deposits, due either to corrosion or iron contamination in the water, are very dark coloured. Iron deposits in boilers are most often magnetic. They are soluble in hot acid giving a dark brown coloured solution.

If unchecked, scaling causes progressive lowering of the boiler efficiency by heat retardation, acting as an insulator. Eventually, scale built-up will cause the tube to overheat and rupture. Boiler deposits can also cause plugging or partial obstruction of corrosive attack underneath the deposits may occur. In general, boiler deposits can cut operating efficiency, produce boiler damage, cause unscheduled boiler outages, and increase cleaning expense.

The purer the feed water is, the weaker the driving mechanism to form scale. Scale-forming minerals that do enter the boiler can be rendered harmless by internal chemical treatment. A long-established technique is to detach the hardness cations, magnesium and calcium, from the scale forming minerals and to replace them with sodium ions.

Presence of Silica

Silica can vaporize into the steam at operating pressures as low as 28 bars. Its solubility in steam increases with increased temperature; therefore, silica becomes more soluble as steam is superheated. The conditions under which vaporous silica carryover occurs have been thoroughly investigated and documented. Researchers have found that for any given set of boiler conditions using demineralized or evaporated quality make-up water, silica is distribute between the boiler water and the steam in a definite ratio. This ratio depends on two factors: boiler pressure and boiler water pH. The value of the ratio increases almost logarithmically with increasing pressure and decreases with increasing pH. If the silica enters the boiler water, the usual corrective action is to increase boiler blowdown, to decrease it to acceptable levels and then to correct the condition that caused the silica contamination.

Corrosion

Corrosion is the reversion of a metal to its ore form. Iron, for example, reverts to iron oxide as the result of corrosion. The process of corrosion however is a complex electro chemical reaction and it takes many forms. Corrosion may produce general attach over a large metal surface or it may result in pinpoint penetration of metal. Corrosion is a relevant problem caused by water in boilers. Corrosion can be of widely varying origin and nature due to the action of dissolved oxygen, to corrosion currents set up as a result of heterogeneities on metal surfaces, or to the iron being directly attacked by the water.

While basic corrosion in boilers may be primarily due to reaction of the metal with oxygen, other factors such as stresses, acid conditions, and specific chemical corrodents may have an important influence and produce different forms of attack. It is necessary to consider the quantity of the various harmful substances that can be allowed in the boiler water without risk of damage to the boiler. Corrosion may occur in the feed-water system as a result of low pH water and the presence of dissolved oxygen and carbon dioxide.

Starting form these figures and allowing the amount that can be blown down, the permitted concentration in the make-up water is thus defined.

Corrosion is caused principally by complex oxide-slag with low melting points. High temperature corrosion can proceed only if the corroding deposit is in the liquid phase and the liquid is in direct contact with the metal. Deposits also promote the transport of oxygen to the metal surface.

Corrosion in the boiler proper generally occurs when the boiler water alkalinity is low or when the metal is exposed to oxygen bearing water either during operation or idle periods. High temperatures and stresses in the boiler metal tend to accelerate the corrosive mechanisms. In the steam and condensate system corrosion is generally the result of contamination with carbon dioxide and oxygen. Specific contaminants such as ammonia or sulphur bearing gases may increase attack on copper alloys in the system.

Cracking in boiler metal may occur by two different mechanisms. In the first mechanism, cyclic stresses are created by rapid heating and cooling and are concentrated at points where corrosion has roughened or pitted the metal surface. This is usually associated with improper corrosion prevention. The second type of corrosion fatigue cracking occurs in boilers with properly treated water. In these cases corrosion fatigue is probably a misnomer. These cracks often originate where a dense protective oxide film covers the metal surfaces and cracking occurs from the action of applied cyclic stresses. Corrosion fatigue cracks are usually thick, blunt and cross the metal grains. They usually start at internal tube surfaces and are most often circumferential on the tube.

Corrosion control techniques vary according to the type of corrosion encountered. Major methods include maintenance of the proper pH, control of oxygen, control of deposits, and reduction of stresses trough design and operational practices. Deaeration and recently the use of membrane contractors are the best and most diffused ways to avoid corrosion removing the dissolved gasses (mainly O_2 and CO_2).

Foaming and Priming

Boiler water carry-over is the contamination of the steam with boiler-water solids. Bubbles or froth actually build up on the surface of the boiler water and pass out with the steam. This is called foaming and it is caused by high concentration of any solids in the boiler water. It is generally believed, however, that specific substances such as alkalis, oils, fats, greases, certain types of organic matter and suspended solids are particularly conducive to foaming. In theory suspended solids collect in the surface film surrounding a steam bubble and make it tougher. The steam bubble therefore resists breaking and builds up foam. It is believed that the finer the suspended particles the greater their collection in the bubble.

Priming is the carryover of varying amounts of droplets of water in the steam (foam and mist), which lowers the energy efficiency of the steam and leads to the deposit of salt crystals on the super heaters and in the turbines. Priming may be caused by improper construction of boiler, excessive ratings, or sudden fluctuations in steam demand. Priming is sometimes aggravated by impurities in the boiler-water.

Some mechanical entertainment of minute drops of boiler water in the steam always occurs. When this boiler water carryover is excessive, steam-carried solids produce turbine blade deposits. The accumulations have a composition similar to that of the dissolved solids in the boiler water. Priming is common cause of high levels of boiler water carryover. These conditions often lead to super heater tube failures as well. Priming is related to the viscosity of the water and its tendency to foam. These properties are governed by alkalinity, the presence of certain organic substances and by total salinity or TDS. The degree of priming also depends on the design of the boiler and its steaming rate.

The most common measure to prevent foaming and priming is to maintain the concentration of solids in the boiler water at reasonably low levels. Avoiding high water levels, excessive boiler loads, and sudden load changes also helps. Very often contaminated condensate returned to the boiler system causes carry-over problems. In these cases the condensate should be temporarily wasted until the source of contamination is found and eliminated. The use of chemical anti-foaming and anti-priming agents, mixtures of surface-active agents that modify the surface tension of a liquid, remove foam and prevent the carry-over of fine water particles in the stream, can be very effective in preventing carry-over due to high concentrations of impurities in the boiler-water.

Desalination of Water

Due to its high cost, energy intensiveness and overall ecological footprint, most environmental advocates view desalinization (or desalination), the conversion of salty ocean water into fresh water, as a last resort for providing fresh water to needy populations. Sourcing fresh water from streams, rivers, lakes and underground aquifers and adhering to strict water conservation measures are much more viable for both economic and environmental reasons in most situations, although some desert regions with thirsty and growing populations may not have many such options.

The relationship between desalinization and climate change is complex. Global warming has increased droughts around the world and turned formerly verdant landscapes into near deserts. Some long held fresh water sources are simply no longer reliably available to hundreds of millions of people around the world.

Meanwhile, expanding populations in desert areas are putting intense pressure on existing fresh water supplies, forcing communities to turn to desalinization as the most expedient way to satisfy their collective thirst. But the process of desalinization burns up many more fossil fuels than sourcing the equivalent amount of fresh water from fresh water bodies. As such, the very proliferation of desalinization plants around the world, some 13,000 already supply fresh water in 120 nations, primarily in the Middle East, North Africa and Caribbean, is both a reaction to and one of many contributors to global warming. Beyond the links to climate problems, marine biologists warn that widespread desalinization could take a heavy toll on ocean biodiversity; as such facilities' intake pipes essentially vacuum up and inadvertently kill millions of plankton, fish eggs, fish larvae and other microbial organisms that constitute the base layer of the marine food chain. And, according to Jeffrey Graham of the Scripps Institute of Oceanography's Centre for Marine Biotechnology and Biomedicine, the salty sludge leftover after desalinization for every gallon of freshwater produced, another gallon of doubly concentrated salt water must be disposed of can wreak havoc on marine ecosystems if dumped willy-nilly offshore. For some desalinization operations, says Graham, it is thought that the disappearance of some organisms from discharge areas may be related to the salty outflow.

Fresh Water from Water Sources

The world's water consumption rate is doubling every 20 years, outpacing by two times the rate of population growth. It is projected that by the year 2025 water demand will exceed supply by 56%, due to persistent regional droughts, shifting of the population to urban coastal cities, and water needed for industrial growth. The supply of fresh water is on the decrease. Water demand for food, industry and people is on the rise. Lack of fresh water reduces economic development and lowers living standards. Clearly, there is a critical worldwide need to better manage this increasingly valuable resource. Desalination systems from GE Water and Process Technologies can make abundant fresh water both from seawater and from challenging brackish sources.

Seawater Desalination

Oceans make up 97% of the world's supply of water. Desalination using seawater reverse osmosis (SWRO) membrane technology has become a viable option for the development of new regional water supplies.

GE Water and Process Technologies is the world leader in the supply of reliable seawater SWRO membrane desalination systems. GE's desalination projects range in size from small 2,000 cubic metre/ day (370 gpm) plants — providing potable water to hotels and resort complexes — all the way up to projects like the 200,000 cubic metre/ day (53 MGD) Hamma Desalination Plant, the largest of its kind in

Africa. Hamma supplies desperately-needed drinking water to over 20% of the population of Algeria's capital city.

Brackish Water Desalination

Brackish water, containing minerals and salts typically less than 5,000 ppm total dissolved solids (TDS), can be economically treated with today's reverse osmosis (RO) and electrodialysis reversal (EDR) systems. Effective mineral and salt removal converts previously unusable waters to high-purity resources for drinking, irrigation or industrial process uses. GE pioneered brackish water desalination in the early 1950s and continues to lead the way withcost-effective, environmentally friendly treatment systems for groundwater and surface water sources. Our global installed capacity includes approximately 250 mgd (950,000 m3/day) of EDR systems and 600 mgd (2.3 million m3/day) of RO systems.

Composition of Water

Water consists of only hydrogen and oxygen. Both elements have natural stable and radioactive isotopes. Due to these isotopes, water molecules of masses roughly 18 ($H_2{}^{16}O$) to 22 ($D_2{}^{18}O$) are expected to form. Isotopes and their abundances of H and O are given below. From these data, we can estimate the relative abundances of all isotopic water molecules.

Abundances (% or halflife) of hydrogen and oxygen isotopes

H	^{2}D	^{3}T		
99.985%	0.015%	12.33 y		
^{14}O	^{15}O	^{16}O	^{17}O	^{18}O
70.6 s	122 s	99.762%	0.038%	0.200%

Relative abundance of isotopic water

$H_2{}^{16}O$	$H_2{}^{18}O$	$H_2{}^{17}O$	$HD^{16}O$	$D_2{}^{16}O$	$HT^{16}O$
99.78%	0.20%	0.03%	0.0149%	0.022 ppm	trace
18	20	19	19	20	20 amu

The predominant water molecules $H_2{}^{16}O$ have a mass of 18 amu, but molecules with mass 19 and 20 occur significantly. Because the isotopic abundances are not always the same due to their astronomical origin, The isotopic distribution of water molecules depends on its source and age. Its study is linked to other sciences.

In particular, $D_2{}^{16}O$ is called heavy water, and it is produced by enrichment from natural water. Properties of heavy water are particularly interesting due to its application in nuclear technology.

Structure and Bonding of the Water Molecule

Pure water, H_2O, has a unique molecular structure. The O-H bondlengths are 0.096 nm and the H-O-H angle = 104.5°. This strange geometry can be explained by various methods.

From carbon to neon, the numbers of valence electrons increase from 4 to 8. These elements require 4, 3, 2, 1, and 0 H atoms to share electrons in order to complete the octet requirement. Their Lewis dot structures are shown on the right, and note the trend in bondlengths.

There are six valance electrons on the oxygen, and one each from the hydrogen atom in the water molecule. The eight electrons form two H-O bonds, and left two lone pairs. The long pairs and bonds stay away from each other and they extend towards the corners of a tetrahedron. Such an ideal structure should give H-O-H bond angle of 109.5°, but the lone pairs repel each other more than they repel the O-H bonds. Thus, the O-H bonds are pushed closer, making the H-O-H angle less than 109°.

After the introduction of quantum mechanics, the electronic configuration for the valence electron of oxygen are $2s^2\ 2p^4$. Since the energy levels of $2s$ and $2p$ are close, valence electrons have characters of both s and p. The mixture is called sp^3 hybridization. These hybridized orbitals are shown on the right. The structures of CH_4, NH_3, and H_2O can all explained by these hybrid orbitals of the central atoms. The above approach is the valence bond theory, and both the C-H bonds and lone electron pairs are counted as VSPER pairs in the Valence-shell Electron-Pair Repulsion (VSEPR) model, according to which, the four groups point to the corners of a tetrahedron.

For triatomic molecules such as water, molecular orbital (MO) approach can also be applied to discuss the bonding. The result however is similar to the valence bond approach, but the MO theory gives the energy levels of the electron for further exploration.

Molecular Vibration of Water

Atoms in a molecule are never at rest, and for each type of molecule, there are some normal vibration modes. For the water molecule, the three normal modes of vibrations are symmetric stretching, bending and assymmetric stretching. The vibrations are quantized, as do any microscopic system, and their quantum numbers are designated as υ_1, υ_2 and υ_3. The observed transition bands of D_2O, H_2O, and HDO are given in the table on the right.

The ideal transition bands are centered in the given wavenumbers. However, these wavenumbers are calculated based on isolated molecules with no interaction with any neighbour. When molecules interact with each other, the energy levels are modified, and the bands shift. Many more less intense absorption bands extend into the green part of the visible spectrum. The absorption spectrum of water may contribute to the blue color for lake, river and ocean waters.

Symmetry of Water Molecules

The water molecules are rather symmetric in that there are two mirror planes of symmetry, one containing all three atoms and one perpendicular to the plane passing through the bisector of the H-O-H angle. Furthermore, if the molecules are rotated 180° (360°/2) the shape of the molecule is unperturbed. This indicates that the molecules have a 2-fold rotation axis. The three symmetry elements are 2-fold rotation, and two mirror planes. Both mirror planes contain the rotation axis, and this type of symmetry belongs to the point group $C_{2\upsilon}$.

A point group has a definite number of symmetry elements arranged in certain fashion. Molecules can be classified according to their point groups. Molecules of the same point group have similar spectroscopic characters. Other molecules of $C_{2\upsilon}$ point group are $CH_2{=}O$, CH_2Cl_2, the bent O_3 etc.

Formation of Hydrogen Bonding

Under certain conditions, an atom of hydrogen is attracted by rather strong forces to two atoms instead of only one, so that it may be considered to be acting as a bond between them. This is called hydrogen bond. This statement is from Linus Pauling (1939) in his book *The Nature of the Chemical Bond.* He gave the ion $[F{:}H{:}F]^-$ as an example. At that time, the hydrogen bond was recognized as mainly ionic in nature. The energy associated with hydrogen bond is 8 to 40 kJ/mol.

Normally, the melting point and boiling point of a substance increase with molecular mass. For example the melting points of inert gases are 0.95, 24.48, 83.8, and 116.6 K respectively for He, Ne, Ar, and Kr.

In this table, the melting and boiling points for water are particular high for its small molecular mass. This is usually attributed to the formation of hydrogen bonds. The small electronegative atoms F, O and N are somewhat negatively charged when they are bonded to hydrogen atoms. The negative charges on F, O and N attract the slightly positive hydrogen atoms, forming a strong interaction called hydrogen bond.

Based on the observed absorption at 3546 and 3691 cm^{-1}, Van Thiel, Becker, and Pinmentel (1957, *J. Chem. Phys.* 27 386) suggested the formation of water dimer when trapped in a matrix of nitrogen. Due to hydrogen bonding, water molecules form dimers, trimers, polymers, and clusters. The hydrogen bonds are not necessarily liner.

Structure of Ice

Ice occurs in many places, including the Antarctic. If all the ice melted, the water level of the oceans will rise about 70 m. The structure of ice and the caption are from this link. The density of ice is dramatically smaller than that of water, due to the regular arrangement of water molecule via hydrogen bonds. In an idealized structure of ice, every hydrogen atom is involved in hydrogen bond. Every oxygen atom is surrounded by four hydrogen bonds.

This diagram from caltech.edu, shows the structure of hexagonal ice in (a) and cubic ice in (b). A rod here represents a hydrogen bond. Since the hydrogen bonds are not linear, the real structure is a little more complicated.

The tetrahedral coordination opens up the space between molecules. On each hydrogen bond, shown by a rod joining the oxygen atoms, lies one proton in an asymmetric position (not shown). Bond lengths, 275 pm, are indicated. Ordinary ice is hexagonal. and the hexagonal *c* axis is labelled 732 pm, and one of the hexagonal *a* axes is labelled 450 pm. If water vapour condenses on very cold substrate at 143-193 K (-130 to -80°C) a cubic phase is formed. In (b) the cubic unit cell is outlined with dashed lines; dimensions are in pm determined at 110 K.

These diagrams can also be used to represent the two forms of diamond, and in this case, the rods joining the atoms represent C-C bonds. Each C-C bondlength is 154 pm. Silicon and germanium crystals have the same structure, but their bondlengths are longer.

The two diamond types of structure are related to the packing of spheres. The hexagonal type has the ABABAB... sequence, whereas the cubic type has the ABCABC... sequence. In both cases, half of the tetrahedral sites are occupied by tetrahedrally bonded carbon atoms. Hexagonal diamonds have been observed in meteorites. The four hydrogen bonds around an oxygen atom form a tetrahedron in a fashion found in the two types of diamonds. Thus, ice, diamond, and close packing of spheres are somewhat topologically related.

A phase diagram of water shows 9 different solid phases (ices). Ice Ih is the ordinary ice. In addition to ice Ic from vapour deposition, conditions for nine phases are shown. Aside from ice I, other phases are formed and observed under high pressure generated by machines built by scientists. So far, ten different forms of ice have been observed, and some ice forms exist at very high pressure. The pressure deep under the polar (Antarctic) ice cap is very high, but we are not able to make any direct observation or study.

There is a report of the 11th ice, and the ice phase diagram and drawings of ice structures given here is extremely interesting.

The Autoionization of Water

The Autoionization of Water in the formation of ions according to

$$HOH(l) + HOH(l) = H_3O^+ + OH^-$$

This is an equilibrium process and is characterised by an equilibrium constant, K'_w:

$$K'_w = \frac{[H_3O^+]\ [OH^-]}{[H_2O]}$$

Since $[H_2O]$ = 1000/18 = 55.56 M, and remains rather constant under any circumstance, we usually write

$$K_w = [H_3O^+]\ [OH^-] = 10^{-14} \text{ (or 1e-14)}$$

$$pK_w = -\log K_w \quad \text{(defined)} = 14 \text{ (at 298 K)}$$

For neutral water, $[H_3O^+] = [OH^-]$ = 1e-7 at this temperature. Furthermore, we define

$$pH = -\log[H_3O^+]$$

$$pOH = -\log[OH^-]$$

$$pH = pOH = 7 \text{ at 298 K; (in neutral solutions)}$$

Leveling effect of water and acid-base characters

The strength of strong acids and bases is dominated by the autoionization of water. In aqueous solutions, the strongest acid and base are the hydronium ion, H_3O^+, and the hydroxide ion OH^- respectively. Acids HCl, HBr, HI, HNO_3, $HClO_3$, $HClO_4$, and H_2SO_4 completely ionize in water, making them as strong as H_3O^+ due to the leveling effect of water. Furthermore, strong acids, strong bases, and salts completely ionize in their aqueous solutions. For example, HCl is a stronger acid than H_2O, and the reaction takes place as HCl dissolves in water.

$$HCl + H_2O = Cl^- + H_3O^+$$

A similar equation can be written for another strong acid. On the other hand, a stong base also react with water to give the stong base species, OH^-.

$$H_2O + B^- = OH^- + HB$$

For example, O^{2-}, CH_3O^-, and NH_3 are strong bases. The leveling effect also apply to bases.

Amphiprotic Species

Equilibria of acids and bases, are interesting chemistry. When an acid and a base differ by a proton, they are called a conjugate acid-base pair. A water molecule is a weak acid and base, due to its ability to accept or donate a proton. Such properties make water an amphiprotic species. In fact, H_3O^+, H_2O and OH^- are amphiprotic, as are some other conjugate acid-base pairs of weak acids and bases. If several acids and bases are dissolved in water, all equilibria must be considered. To estimate the pH of these solutions requires the exact treatment of several equilibrium constants. For example, many species dissolve in rain water, and many equilibria must be considered. Detail consideration and examples are given in Acid-Base Reactions

Carbon dioxide in the air dissolve in rain water, lakes and rivers. A solution of CO_2 involves the following reaction:

Reaction	*K formula*	*K value*
$H_2O(l) + CO_2(g) = H_2CO_3(l)$	$1/P_{CO2}$	?
$H_2CO_3 = HCO_3^- + H^+$	$[HCO_3^-]\ [H^+]\ /\ [H_2CO_3]$	5e-7
$HCO_3^- = CO_3^{-2} + H^+$	$[CO_3^{-2}]\ [H^+]\ /\ [HCO_3^-]$	5e-11
$HOH(l) + HOH(l) = H_3O^+ + OH^-$	$[H_3O^+]\ [OH^-]$	1e-14

These complicated equilibria make natural water a buffer.

Reactivity of Water Towards Metals

Alkali metals react with water readily. Contact of cesium metal with water causes immediate explosion, and the reactions become slower for potassium, sodium and lithium. Reaction with barium, strontium, calcium are less well known, but they do react readily. Warm water may be needed to react with calcium metal, however. Many metals displace H^+ ions in acidic solutions. This is often seen as a property of acids.

Electrolysis of Water

The *enthalpy of formation* for liquid water, $H_2O(l)$, is -285.830 and that of water vapour is -241.826 kJ/mol. The difference is the heat of vaporization at 298 K. Liquid water and vapour entropies (*S*) are 69.95 and 188.835 kJ K^{-1} mol^{-1} respectively, (*see* Thermodynamic Data. These are entropies, not standard entropies of formation. The entropy of formation for water is obtained by,

$$\Delta S^o_{f\ water} = S^o_{water} - S^o_{H2} - 0.5\ S^o_{O2}$$

$$= 69.95 - 130.68 - 0.5*205.14 \quad \text{(data from Thermodynamic Data)}$$

$$= -163.3\ J\ K^{-1}\ mol^{-1}$$

$$\Delta G^o_{water} = \Delta H - T\ \Delta S \qquad \text{(note } H \text{ in kJ/mol and S in J/mol)}$$

$$\Delta G^o_{water} = -285.83 - 298.15 * 163.3/1000 = -237.13\ kJ$$

The equilibrium constant and Gibb's energy are related,

$$\Delta G^o = -R\ T \ln K$$

$$K = \exp(-\ DG^o\ /\ R\ T) = 3.5e41\ atm^{-3/2}$$

This is a very large value for the formation of water,

$$H_2 + 0.5\ O_2 = 0.5\ H_2O(l).$$

In other words, the reaction is complete, and the possibility of water dissociated into hydrogen and oxygen is very small. A negative value for DG^o indicates an exothermic reaction. The Gibb's energy is the energy released other than pressure-volume work. This redox reaction to form water can be engineered to proceed in a Daniel cell. In this case, the energy is converted into electric energy according to this equation.

$$\Delta G^o_{water} = -n\ F\ E = -237.13\ kJ$$

where n is the number of electrons (= 2) in the redox equation, F is the Faraday constant (= 96485 C), and E is the potential of the Daniel cell. Thus,

$$E = -\frac{-237130J}{2*96485C} = 1.23\ V$$

Ideally, a reverse voltage of 1.23 V is required for the electrolysis of water. But in reality, a little over voltage is required to carry out the electrolysis to decompose water. Furthermore, pure water does not conduct electricity, and acid, base or salt is often added for the electrolysis of water. This link has a demonstration.

Chemistry of Hard and Soft Water

Water Works

Hard water is any water containing an appreciable quantity of dissolved minerals. Soft water is treated water in which the only cation (positively charged ion) is sodium. The minerals in water give it a characteristic taste. Some natural mineral waters are highly sought for their flavor and the health benefits they may confer. Soft water, on the other hand, may taste salty and may not be suitable for drinking.

If soft water tastes bad, then why might you use a water softener? The answer is that extremely hard water may shorten the life of plumbing and lessen the effectiveness of certain cleaning agents. When hard water is heated, the carbonates precipitate out of solution, forming scale in pipes and tea kettles.

In addition to narrowing and potentially clogging the pipes, scale prevents efficient heat transfer, so a water heater with scale will have to use a lot of energy to give you hot water. Soap is less effective in hard water because its reacts to form the calcium or magnesium salt of the organic acid of the soap. These salts are insoluble and form grayish soap scum, but no cleansing lather. Detergents, on the other hand, lather in both hard and soft water. Calcium and magnesium salts of the detergent's organic acids form, but these salts are soluble in water.

Hard water can be softened (have its minerals removed) by treating it with lime or by passing it over an ion exchange resin. The ion exchange resins are complex sodium salts. Water flows over the resin surface, dissolving the sodium. The calcium, magnesium, and other cations precipitate onto the resin surface. Sodium goes into the water, but the other cations stay with the resin. Very hard water will end up tasting saltier than water that had fewer dissolved minerals.

Most of the ions have been removed in soft water, but sodium and various anions (negatively charged ions) still remain. Water can be deionized by using a resin that replaces cations with hydrogen and anions with hydroxide. With this type of resin, the cations stick to the resin and the hydrogen and hydroxide that are released combine to form pure water.

Water is a Chemical H_2O

Water plays an important role as a chemical substance. Its many important functions include being a good solvent for dissolving many solids, serving as an excellent coolant both mechanically and biologically, and acting as a reactant in many chemical reactions. Blood, sweat and tears... all solutions of water.

As chemists we consider water from many perspectives. It is our role to use physical and mathematical laws in application for useful purposes, including diverse perspectives such as living systems, materials and energy. The world of the chemist is a small world-atomic, molecular-which plays a large part in making our lives healthy, comfortable, and hopeful. Because of the diversity of the chemical world, it would be difficult to touch upon all of the applications of water. And for the same reason, it would be impossible to discuss the chemical aspects of water without touching upon the physical, mathematical, and biological aspects of the subject.

Let's start our discussion of water as a chemical with a look at its structure. From a molecular perspective, structure is one of the important features of a substance. Just as you might say that the shape of a key determines its function-which doors it can and cannot open-the structure of a molecule and its composition absolutely determines its functions and properties.

As chemists we have a vested interest not only in understanding how a substance may be used and broken down, but also in knowing how that substance is created. From this perspective let's look at the chemistry which creates water from its elements, hydrogen and oxygen, and the chemistry of water's breakdown, also known as Electrolysis.

Water may be a substance so common that we scarcely make note of it-We waste it, pollute it, let it run down the drain, flush it away... Certainly we take it for granted! However, chemically speaking, water is really not common at all. When compared to other compounds of similar size, composition, and structure-it is absolutely unique! In fact its properties are so unusual that it would be irreplaceable. Let's take a chemical look at these unusual properties, how they arise and what their implications are.

Group	1	2	3	4	5	6	7	8	9	10	11	12	13	14	15	16	17	18
Period 1	1 H																	2 He
2	3 Li	4 Be											5 B	6 C	7 N	8 O	9 F	10 Ne
3	11 Na	12 Mg											13 Al	14 Si	15 P	16 S	17 Cl	18 Ar
4	19 K	20 Ca	21 Sc	22 Ti	23 V	24 Cr	25 Mn	26 Fe	27 Co	28 Ni	29 Cu	30 Zn	31 Ga	32 Ge	33 As	34 Se	35 Br	36 Kr
5	37 Rb	38 Sr	39 Y	40 Zr	41 Nb	42 Mo	43 Tc	44 Ru	45 Rh	46 Pd	47 Ag	48 Cd	49 In	50 Sn	51 Sb	52 Te	53 I	54 Xe
6	55 Cs	56 Ba	57 La	72 Hf	73 Ta	74 W	75 Re	76 Os	77 Ir	78 Pt	79 Au	80 Hg	81 Tl	82 Pb	83 Bi	84 Po	85 At	86 Rn
7	87 Fr	88 Ra	89 Ac	104 Rf	105 Db	106 Sg	107 Bh	108 Hs	109 Mt	110 Ds	111 Rg	112 Cp						
			58 Ce	59 Pr	60 Nd	61 Pm	62 Sm	63 Eu	64 Gd	65 Tb	66 Dy	67 Ho	68 Er	69 Tm	70 Yb	71 Lu		
			90 Th	91 Pa	92 U	93 Np	94 Pu	95 Am	96 Cm	97 Bk	98 Cf	99 Es	100 Fm	101 Md	102 No	103 Lr		

Periodic Table Key

X Synthetic Elements	X Liquids or melt at close to room temp.	X Solids	X Gases	Alkali Metals	Alkali Earth Metals	Transition Metals	Other Metals	Metalloids	Other Non Metals	Halogens	Noble Gases	Rare Earth Metals

Periodic Table

The first periodic table was devised by Dmitri Mendeleev and published in 1869.

Mendeleev found he could arrange the 65 elements that were then known in a grid or table so that each element had:

1. A higher atomic weight than the one on its left.
2. Similar chemical properties to other elements in the same column.

He realized that the table in front of him lay at the very heart of chemistry. In his table he noted gaps-spaces where elements should be but none had yet been discovered.

In fact, just as Adams and Le Verrier could be said to have discovered the planet Neptune on paper, Mendeleev could be said to have discovered germanium (which he called *eka*-silicon because he observed a gap between silicon and tin), gallium (*eka*-aluminum) and scandium (*eka*-boron) on paper, for he predicted their existence and their properties before their actual discoveries.

Although Mendeleev had made a crucial breakthrough, he made little further progress because the Rutherford-Bohr model of the atom had not yet been formulated.

In 1913, Henry Moseley, who worked with Rutherford, showed it is atomic number (charge) and not (as Mendeleev had proposed) atomic weight that is most fundamental to the chemical properties of any element. Like Mendeleev, Moseley was able to predict correctly the existence of new elements based on his work.

And today the elements are still arranged in order of increasing atomic number (Z) as you go from left to right across the table. We call the horizontal rows periods and the vertical rows groups.

We also know an element's chemistry is determined by the way its electrons are arranged-its electron configuration.

The noble gases are found in group 18, on the far right of each period. The reluctance of the noble gases to undergo chemical reactions indicates that the atoms of these gases strongly prefer their own electron configurations-featuring a full outer shell of electrons-to any other.

In contrast to the noble gases, the elements with the highest reactivity are those with the greatest need to gain or lose electrons in order to achieve a full outer shell of electrons. Elements sitting in the same group (e.g. the alkali metals in Group 1) all have the same number of outer electrons, leading to similar chemical properties.

Likewise the halogens in Group 17 also have similar properties to one another. When halogens react, they gain an electron to form negatively charged ions. Each ion has the same electron configuration as the noble gas in the same period. The ions are therefore more chemically stable than the elements from which they formed.

There is a progression from metals to non-metals across each period.

The block of elements in groups 3-12 contains the transition metals. These are similar to one another in many ways; they produce colored compounds, have variable valency and are often used as catalysts. The rare earth elements can be divided into lanthanides (elements 58-71) and actinides (elements 90-103). The naturally occurring rare earths are found on earth in only very small amounts. The actinides include most of the well-known elements that take part in or are produced by nuclear reactions. No element with atomic number higher than 92 occurs naturally. These elements are produced artificially in nuclear reactors or particle accelerators.

Basic Water Chemistry

The Properties of Water

Water is the most common substance known to man, as well as the most important. In vapour, liquid or solid form, water covers more than seventy percent of the Earth's surface, and is a major component of the atmosphere. Water is also an essential requirement for all forms of life. Most living things are largely made up of water. Human beings, for example, consist of about two-thirds water.

Pure water is a clear, colorless, and odorless liquid that is made up of one oxygen and two hydrogen atoms. The chemical formula of the water molecule, H_20, was defined in 1860 by the Italian scientist Stanislao Cannizzarro. Water is a very powerful substance that acts as a medium for many reactions, which is why it is often referred to as the "universal solvent." Although pure water is a poor conductor of electricity, impurities that occur naturally in water transform it into a relatively good conductor. Water has unusually high boiling (100° C/212° F) and freezing (0° C/32° F) points. It also shows unusual volume changes with temperature. As water cools, it contracts to a maximum density of 1 grain per cubic centimeter at 4° C (39° F). Further cooling actually causes it to expand, especially when it reaches the freezing point. The fact that water is denser in the liquid form than the solid form explains why an ice cube floats in a beverage, or why a body of water freezes from the top down. While the density property of water is of little importance to the beverage example, it has a tremendous impact on the survival of aquatic life inhabiting a body of water.

Pond Water Chemistry

This is a composite of a series of articles dealing with the chemical makeup of pond water. How to measure what is in it, what is good, what is bad, and what to do about it. Before starting, I would like to discuss the name applied to what has typically been called the "bio-filter." A filter is defined as a porus device through which water (or gas) is passed to separate out matter in suspension. The biologic activity within the pond "filter" does not trap the matter in suspension but acts on dissolved components that could not be separated regardless of how fine the filter pores. Although this device may perform a dual role as a mechanical filter, to emphasize the processes of interest, you will see that I will refer it as the biologic converter or bio-converter, not as a filter.

By introducing fish into your pond, you have assumed the responsibility for the care of these creatures. This includes not only feeding them but also providing them with a healthy environment in which they can live and thrive. Partial determination of the quality of this liquid environment can be made through chemical measurements. It seems somewhat ludicrous that some one would spend hundreds or thousands of dollars to build a pond and then add hundreds or thousands of dollars of beautiful Koi but would not buy and learn how to use a ten dollar Nitrite test kit. This doesn't mean that one must test the water every few minutes or even every few days. An established pond with the fish appearing healthy should be checked every month or so. It is only when something out of the ordinary is observed and possibly during seasonal changes when an additional test or two might be needed. A simple test at the right time may prevent a small problem from becoming a catastrophe. When starting up a new pond or bio-converter system, daily tests may be required until the converter comes on line, then weekly for a couple of months until the system has stabilized.

Test Kits in Estimated Order of Importance;

(Required):

1. pH
2. Ammonia
3. Nitrite
4. Thermometer (Nice To Have)
5. Alkalinity
6. Salinity
7. Nitrate
8. Dissolved Oxygen
9. Chlorine
10. Hardness

Just as when medicinally treating your pond, it is imperative to know the total amount of water in your pond and converter/filter system as accurately as possible. Over treatment or under treatment with chemicals can be equally disastrous. Don't guess on quantities, measure them!

Do not confuse the terms water quality and water clarity. Crystal clear water can contain compounds that are deadly to your fish. Green water, caused by excessive phyto plankton growth can actually be beneficial to the fish although not very beneficial to the pond keeper who can't see them. Water clarity can give some indications as to mechanical filtration effectiveness.

A pond with a biologic converter and filled with Koi is a rather complex, somewhat self-con tained ecological system. Each component of this system requires the other components to survive and prosper. The basic portion of the cycle is shown above. Fish waste and other organic waste is converted by bacteria and fungi to Ammonia compounds. These compounds can be injurious to the fish, but a healthy biologic converter populated with Nitrosomonas bacteria consume these Ammonia compounds and convert them to Nitrites. Unfortunately, the Nitrites are just as toxic to the fish as the Ammonia. Again, the biologic converter comes into play with a population of Nitrobacter bacteria that convert the Nitrites to Nitrates. The Nitrates are basically inert to the fish but usable by plants and algae within the pond. As the plants and algae grow and the Koi eat them, the cycle starts all over again. The Nitrosomonas and Nitrobacter bacteria are called aerobic since they require Oxygen to convert their "food" to energy just like the fish.

Ammonia

Ammonia, NH3, measured in parts per million (ppm), is the first measurement to determine the "health" of the biologic converter. Ammonia should not be detectable in a pond with a "healthy" bio-converter. The ideal and normal measurement of Ammonia is zero. When ammonia is dissolved in water, it is partially ionized depending upon the pH and temperature. The ionized ammonia is called Ammonium and is not toxic to the fish. As the pH drops and the temperature decreases, the ionization and Ammonium increases which decreases the toxicity. As a general guideline for a water temperature of 70°F., most Koi would be expected to tolerate an Ammonia level of 1 ppm if the pH was 7.0, or even as high as 10.0 if the pH was 6.0. At a pH of 8.0, just 0.1 ppm could be dangerous.

Test kits are available in two basic types. Both read the total of Ammonia and Ammonium, so without knowing the temperature and pH, the toxicity cannot be determined. Suffice it to say that the only good Ammonia reading is zero. The Nessler method type test normally uses drops with a colormetric chart. The Nessler test detects both free Ammonia/Ammonium and also that chemically bound with anti-Ammonia chemical treatments (more about these later). The Salicylate type test is a dual step, using liquid, pill or powder also with an associated color chart. It takes longer to perform and measures only the free Ammonia/Ammonium. Since only the free Ammonia is harmful to the fish, the Nessler test can be misleading under certain conditions but provides additional information under others. The recommended test kit should be able to detect 0-1 ppm of Ammonia particularly for ponds with normal pH levels above 7.0. A wider range kit, 0-5 ppm, would also be useful, particularly for those ponds with a typical pH of under 7.0. An Ammonia test kit is considered to be a requirement for all pond keepers.

Effects

Ammonia tends to block oxygen transfer from the gills to the blood and can cause both immediate and long term gill damage. The mucous producing membranes can be destroyed, reducing both the external slime coat and damaging the internal intestinal surfaces. Fish suffering from Ammonia posioning usually appear sluggish, often at the surface as if gasping for air.

Source

Ammonia is a gas primarily released from the fish gills as a metabolic waste from protein breakdown, with some lesser secondary sources such as bacterial action on solid wastes and urea.

Control

Ammonia is removed by bacterial action in the bio-converter and some is directly assimilated by the algae in the pond. Nitrosonomas bacteria consume the Ammonia and produce Nitrites as a waste product. A significant portion of this bacterial action can occur on the walls of the pond as well as in the bio-converter. Ammonia readings may increase with a sudden increase in bio-converter load until the bacterial colony grows to accept the added material. This can happen following the addition of a large number of new fish to a pond or during the spring as the water temperature increases. Fish activity can often increase faster following a temperature increase than the bacterial action does. A bio-converter that becomes partially obstructed with waste and/or develops channels through the media may operate at a reduced effectiveness that can also cause the Ammona levels to increase.

Treatment

Chemical treatments to counteract Ammonia toxcity are available commercially under various trade names. These treatments, most of which are based on Formaldehyde, form a chemical bond with the Ammonia that prevents it from being harmful to the fish. They do not remove it from the pond. The bio converter does the actual removal. Although most of these products use a dosage of 50 ml per 100 gallons to chemically bind up to 1 ppm of Ammonia, be sure and check the manufacturer's directions before use. Note that Nessler type test kits will still show chemically bound Ammonia to be present until the bio-converter bacteria actually consume it. If a pond has a healthy bio-converter, there is not only no need to treat with Ammonia binding chemical agents, it is better not to use them at all.

When Ammonia is detected (assuming a pH of about 7.5):

1. Increase aeration to maximum. Add supplemental air if possible.
2. Stop feeding the fish if detected in an established pond, reduce amount fed by half if starting up a new bio-converter/pond.
3. Check an established pond bio-converter for probable clean out requirement.
4. For an ammonia level of 0.1 ppm, conduct a 10% water change out. For a level of 1.0 ppm, conduct a 25% change out. CAUTION: If the tap water has a higher pH than that of the pond, adding the replacement water may make the situation worse.
5. Chemically treat for twice the amount of Ammonia measured.
6. Consider transferring fish if the Ammonia level reaches 2.5 ppm.
7. If starting up a new bio-converter/pond, discontinue use of any UV Sterilizers, Ozone Generators, and Foam Fractionators (Protein Skimmers).
8. Retest in 12 to 24 hours.
9. Under Emergency conditions only, consider chemically lowering the pH one-half unit (but not below 6.0).

Nitrite

Nitrite, NO2-N, measured in parts per million (ppm), is the second chemical measurement made to determine the "health" of the biologic converter. Nitrite should not be detectable in a pond with a properly functioning bio-converter. Thus the ideal and normal measurement of Nitrite is zero. A low Nitrite reading combined with a significant Ammonia reading indicates the Ammonia- Nitrite biologic converter action is not established, while a low Ammonia reading with a detect able Nitrite reading indicates that the Nitrite-Nitrate bacterial conversion activity is not yet working. Test kits are available in pill, powder, or droplet forms with color charts. Recom mended test kit range 0-4 ppm. A Nitrite test kit is considered to be a requirement for all pond keepers.

Source

Nitrite is produced by the autotrophic Nitrosomonas bacteria combining Oxygen and Ammonia in the bio-converter and to a lesser degree on the walls of the pond. Just as with Ammonia, Nitrite readings may increase with a sudden increase in bio-converter load until the bacterial colony grows to accept the added material. This can happen following the addition of a large number of new fish to a pond or during the spring as the water temperature increases. Fish activity can often increase faster following a temperature increase than the bacterial action does. A bio converter that becomes partially obstructed with waste and/or develops channels through the media may operate at a reduced effectiveness that can also cause the Nitrite levels to increase.

Effects

Nitrite has been termed the invisible killer. The pond water may look great but Nitrite cannot be seen. It can be deadly, particularly to the smaller fish, in concentrations as low as 0.25 ppm. Nitrite damages the nervous system, liver, spleen, and kidneys of the fish. Even lower concentrations over extended periods can cause long term damage. Short term, high intensity, “spikes” which often occur during bio-converter startup may go undetected yet cause problems to develop within the fish months later. A common indication of a fish that has endured a Nitrite spike in the past is that the gill covers may be slightly rolled outward at the edges. They do not close flat against fish’s body.

Control

About the only control of Nitrite is through the maintenance of a “healthy” bio-converter. Within the media, Nitrobacter bacteria combine Oxygen with the Nitrite to convert it to the relatively benign Nitrate. The Nitrobacter bacteria receive considerably less energy from this conversion process than do the Nitrosomonas bacteria in the Ammonia to Nitrite process. For this reason, they are not as hardy and tend to be the last to come and the first to go when a problem occurs within the bio-converter. Water change outs can reduce the levels temporarily by the same amount as the percentage of water changed. The addition of salt helps reduce the toxic effects significantly but should only be used as a interim measure, not as an ongoing treatment.

Whenever 0.25 ppm of Nitrite or more is detected in a pond:

1. Increase aeration to maximum. For a Nitrite level of 1 ppm or greater, add supplemental air, if possible.
2. Stop feeding the fish if detected in an established pond, reduce amount being fed by half if starting up a new bio-converter/ pond.
3. Discontinue use of any UV Sterilizers, Ozone Generators, and Foam Fractionators (Protein Skimmers).
4. For a Nitrite level less than 1 ppm, conduct a 10% water change out and add 1 pound of salt per hundred gallons of changed water.
5. For a level between 1 and 2 ppm, conduct a 25% water change out and add 2 pounds of salt per hundred gallons of changed water.
6. For a level greater than 2 ppm, conduct a 50% water change out and add 3 pounds of salt per hundred gallons of changed water.
7. Retest and repeat above in 24 hours.
8. For Nitrite levels of 4.0 or greater, consider transferring fish.

Nitrate

Nitrate, NO3-N, measured in ppm, is the third and last measurement used to determine the "health" of the bio-converter. Nitrate is produced by the autotrophic Nitrobacter bacteria combining Oxygen and Nitrite in the bio-converter and to a lesser degree on the walls of the pond. A zero Nitrate reading, combined with a non-zero Nitrite reading, indicates the Nitrite-Nitrate bacterial converter action is not established. Test kits are available with dual droplet or pill form with color charts. The recommended test kit range 0-200 ppm. A Nitrate test kit is considered nice to have but not required for the average pond. In an established pond with part of the routine maintenance including 5% to 10% water change outs every two to four weeks, Nitrate levels will normally stabilize in the 50-100 ppm range.

Concentrations from zero to 200 ppm are acceptable. Where Ammonia and Nitrite were toxic to the fish, Nitrate is essentially harmless. There have been reports that high nitrate levels may weaken the colors in Koi but there have also been reports that high nitrate levels can enhance the colors.

Similarly, I have read reports, fortunately not in the same article, that high nitrate levels will both stimulate and suppress spawning activity. If the Nitrate concentration gets too high, the Nitrite-Nitrate converting bacteria (Nitrobacter) may not be able to do their job effectively resulting in a raised Nitrite level. Nitrate is the end result of the nitrification cycle and is very important to plants in their life cycle. This is why the plants in your garden can flourish from being watered with the waste water from your pond (assuming you haven't added too much salt).

Note the large difference in the ranges of the test kits being used to measure Nitrate (200 ppm) as opposed to those for Ammonia and Nitrite (1-4 ppm). Assuming our the bio-converter was converting the equivalent of 1 ppm of Ammonia to the equivalent of 1 ppm of Nitrite to the equivalent of 1 ppm of Nitrate per day, it would take 100 days or over three months, (longer with any water change outs), for the nitrate levels to build up to the 100 ppm level. The Nitrate concentration is controlled naturally through routine water change outs and to a lesser degree through plant/algae consumption.

Temperature

Whether you measure your pond's temperature in degrees Centigrade or degrees Fahrenheit or both, a thermometer is considered a requirement for all ponds. A floating pool or spa thermometer is good. It is recommended that it be floated in the filter/converter system or tied to an easy access point at the edge of the pond. At a slightly higher cost, the electronic indoor/outdoor thermometers on the market (i.e. Radio Shack) provide a continuous digital readout. Just drop the end of the waterproof outdoor probe into the water. (Note: Small floating glass aquarium thermometers have been swallowed by Koi.)

Temperature Ideal Range 65°F-75°F (20°C-25°C)

Acceptable 35°F-85°F (2°C-30°C)

The temperature of the pond normally follows that of its surroundings although with a delay related to the size of the pond.

Direct exposure of the pond to open sky can cause larger swings in temperature. Direct sunlight during the day can cause the temperature to rise higher and heat loss on clear nights can cause the temperature to drop lower than shaded ponds. A clear night sky can absorb a large amount of heat from a small pond and actually drive the pond temperature below air temperature.

Events generally happen faster at higher temperatures and in smaller ponds. Over normal temperature ranges, biologic activity doubles for each 10° rise in temperature. The toxicity of Ammonia increases as the temperature rises and the amount of Dissolved Oxygen that the water can hold decreases. Although Koi have been known to survive for limited periods at 100°F and even higher, the mortality rate of fish conditioned to 75°F water increases rapidly above 85°F. Above 80°F, supplemental air may be required. Below 55°F (12°C), Koi stop producing antibodies and at about 45°F (7°C) enter a state similar to hibernation. Bio-converter bacteria reproduction ceases at about 40°F (5°C).

Feeding fish versus temperature.

Less than 50°F........Do Not Feed

50°F-60°F.............2-4 times weekly

60°F-85°F.............2-4 times daily

Above 85°F............Do Not Feed

In all cases, try to feed only what the fish will normally consume in 10 minutes. Remove any uneaten food within an hour.

Fish do not like changes in their environment of any kind including temperature. Any changes add stress to the fish and the larger and faster the changes, the greater the stress. This is considered by many to be the primary reason that fish do better in larger ponds. Another time that the Koi are subjected to stress from temperature changes is when they are being added to a pond from another location. My recommendation is that if the fish have been bagged for more than two hours, it is better to release them immediately than to subject the fish to the "bad" water in the bag for an additional half-hour. Thirty minutes of floating will prevent a sudden shock if the temperature difference is large, but it will not acclimatize the fish to the new temperature. Actual temperature acclimation of a fish takes several days, similar to us dealing with jet lag. It is not only the temperature the fish needs to be accustomed to but also the pH, hardness, alkalinity, “the taste”, etc. of it’s new surroundings. Other than providing some shade (summer and winter), little can be done or normally needs to be done to control an outdoor pond’s temperature.

Agricultural Chemistry

Agricultural chemistry is the study of both chemistry and biochemistry which are important in agricultural production, the processing of raw products into foods and beverages, and in environmental monitoring and remediation. These studies emphasize the relationships between plants, animals and bacteria and their environment.

Agricultural chemistry often aims at preserving or increasing the fertility of soil, maintaining or improving the agricultural yield, and improving the quality of the crop.

History

In 1815, Humphry Davy publishes “*Elements of Agricultural Chemistry*”, (one of?) the first scientific study on the subject.

In 1840, Justus von Liebig publishes “*Chemistry in its application to agriculture and physiology*”.

Introduction

Modern agriculture depends quite heavily on the advances that have been made in science, and chemistry in particular, to maximize the yield of crops and animal products.Fertilizers, pesticides, and antibiotics play ever increasing roles in this field.

Fertilizers are perhaps the most widely used form of chemical in agriculture. Fertilizers are added to the soil in which crops are growing to provide nutrients required by the plants. Fertilizers can be divided into two categories: organic and inorganic. Organic fertilizers are derived from living systems and include animal manure, guano (bird or bat excrement), fish and bone meal, and compost.

These organic fertilizers are decomposed by microorganisms in the soil to release their nutrients. These nutrients are then taken up by the plants. Inorganic or chemical fertilizers are less chemically complex and usually more highly concentrated. They can be formulated to provide the correct balance of nutrients for the specific crop that is being grown. Both organic and inorganic fertilizers supply the nutrients required for maximum growth of the crop. Inorganic fertilizers contain higher concentrations of chemicals that may be in short supply in the soil. The major or macro- nutrients in inorganic fertilizers are nitrogen, phosphorous, and potassium. These fertilizers also may provide other nutrients in much smaller quantities (micro-nutrients).

With the expansion of cities due to increases in population, there has been a loss of agricultural land. Appropriate use of fertilizers to increase crop yield has in part counterbalanced this loss of land. The use of fertilizers is not without controversy, however. There are concerns that adding supplements of nitrogen, particularly in the form of inorganic fertilizers, can be detrimental. It is thought that adding additional nitrogen to the soil can disrupt the action of nitrogen-fixing bacteria, an important part of the nitrogen cycle.

If these nitrogen-fixing bacteria in the soil are killed, then less nitrogen is added naturally. As a consequence, more and more fertilizer must be applied. Inorganic nitrogen fertilizers are relatively cheap and they are often added to arable land in excessive amounts. The crop does not assimilate all of this extra nitrogen, but instead the nitrogen can run off the land and enter the water supply. High levels of nitrogen in water can lead to eutrophication which can trigger algal and bacterial blooms. These organisms remove oxygen from the water faster than it is replaced by diffusion and photosynthesis, causing some other aquatic animals to die from oxygen deprivation. High levels of nitrate in drinking water, which can be due to agricultural runoff, have been implicated in human health problems, such as blue baby syndrome (methemoglobinemia).

Pesticides are another important group of agricultural chemicals. They are used to kill any undesired organism interfering with agricultural production. Pesticides can be divided into fungicides, herbicides, and insecticides. Fungicides are used to control infestations of fungi, and they are generally made from sulfur compounds or heavy metal compounds. Fungicides are used primarily to control the growth of fungi on seeds. They are also used on mature crops, although fungal infestation are harder to control at this later stage.

Herbicides are weed killers that are used to destroy unwanted plants. Generally herbicides are very selective, since they would be useless for most applications if they were not. A general non-selective herbicide can be used to clear all plants from a particular area. However, appropriate treatment must be carried out to remove the herbicide or render it ineffective if that area is to be used for subsequent plant growth. Herbicides can be used to kill weeds that grow among crops and reduce the value of the harvest. They can also be used to kill plants that grow in fields used for grazing by animals, since some plants can be poisonous to livestock or can add unpleasant flavors to the meat or milk obtained from the livestock. Breeding and genetic manipulation are used to introduce herbicide resistance to crops, allowing the use of more broad-spectrum herbicides that can kill more weed species with a single application. Herbicides include a wide range of compounds, such as common salt, sulfates, and ammonium and potassium salts. In the 1940s 2,4-D (2,4 trichlorophenoxyacetic acid) was developed and this herbicide is still widely used today. The use of a related compound, 2,4,5-T (2,4,5 trichlorophenoxyacetic acid), is now controlled because of its potentially harmful effects. 2,4,5-T was a constituent of Agent Orange, a defoliant used during the Vietnam War.

Insecticides are chemicals that are used to kill insect pests. Insects can spread livestock diseases, can eat stored grain, and can feed on growing crops. Not all insects are harmful, and certain species of insects are needed to pollinate plants to ensure that they set seed. Many insecticides are non-selective and kill all insects, beneficial as well as harmful. Some insecticides, which are very effective at killing insects, have other problems associated with them. For example, DDT (dichlorodiphenyltrichloroethane) persists in the environment and is concentrated in the food chain. With high levels of exposure, DDT can directly kill fish and birds by paralyzing their nerve centers. In lower concentrations, it can weaken bird's egg shells and cause sharp declines in reproductive rates. Insecticides work in a number of ways. Some are direct poisons (chrysanthemic acids, contact poisons, systemic poisons), while others are attractants or repellents that move the insects to a different location (fumigation acrylonitrile). Some insecticides will only attack a particular stage of an insect's life cycle and this can make them more specific.

Antibiotics and growth hormones are routinely used as feed supplements for a number of animals. These additives are supplied to keep the animals free from disease and to help them grow to a marketable size as quickly as possible. However, the indiscriminate use of antibiotics can cause problems, since this can lead to the development of resistant strains of microorganisms or sensitization to the antibiotic among people who eat these animal products. The effects on humans of eating the meat of animals treated with growth hormones are poorly understood at the present time.

Agricultural chemistry has provided us with more and cheaper food than ever before. It has also allowed food to be produced in areas that previously were unsuitable for agriculture. The application of chemicals to farming has been one of the chemical success stories of the twentieth century. This is not to say that there have not been problems, the most famous being DDT. In the 1980s and 1990s there has been a backlash against the application of chemicals to foodstuffs in the western world. This has led to the production of organic and green products that are produced without artificial application of chemicals. These products are often more expensive to produce and this increased price is passed on to the consumer. However, there is much research to suggest that appropriate organic techniques can be competitive in cost with typical chemical agriculture and that Third World countries would benefit greatly from many organic practices.

In addition to the applications outlined above, chemicals also have other agricultural uses. For example, sulfur dioxide can be used to keep grain fresh and useable for a longer period of time than untreated grain. Other chemicals can be added to promote the ripening of fruits or the germination of seeds. It is difficult to estimate the monetary value of agricultural chemicals, but many multi- national corporations are involved in their manufacture and use. Agricultural chemistry has increased the diversity of the human diet and has led to a greater overall availability of food, both animal and plant.

Inorganic Components of the Agricultural Ecosystem

Soil is the primary medium in which biological activity and chemical reactions occur. It is a three-phase system consisting of solid, liquid, and gas. Approximately 50 percent of the volume in a typical agricultural soil is solid material classified chemically as either organic or inorganic compounds. Organic materials usually constitute 1 to 5 percent of the weight of the solid phase. The remainder of the soil volume is pore space that is either filled with gases such as CO_2 and O_2, or water.

Surface area and charge characteristics of the inorganic portion of the solid phase control chemical reactivity. Soil particles are classified based on their size, with sand-sized particles having diameters of 2 to 0.05 millimeters (0.08 to 0.002 inches) and silt-sized particles from 0.05 to 0.002 millimeters (0.002 to 0.00008 inches). Clay-sized materials of less than 0.002 millimeters (0.00008 inches) in diameter have the largest surface area per unit weight, reaching as much as 800 meters (2,625 feet) squared per gram. Because of large surface areas, clay-sized materials greatly influence the sorption of chemicals such as fertilizers and pesticides and play a major role in catalyzing reactions.

Crystalline layer silicates or phyllosilicates present in the clay-sized fraction are especially important because they function as ion exchangers. Most phyllosilicates have a net negative charge and thus attract cations. This cation exchange capacity (CEC) controls whether plant nutrients, pesticides, and other charged molecules are retained in soil or if they are transported out of the soil system. In contrast, aluminum and iron oxides also present in the clay-sized fraction typically possess a net positive charge or an anion exchange capacity (AEC). Soils in temperate regions are dominated most often by solid phase materials that impart a net CEC, whereas soils in tropical regions often contain oxides that contribute substantial AEC.

Organic Components of the Agricultural Ecosystem

Organic materials contained within the solid phase, although only a small percentage of the total soil weight, are extremely important in controlling chemical and physical processes in soil. Organic matter exists in the form of recognizable molecules such as proteins and organic acids, and in large polymers called humic materials or humus. Humus is dominated by acidic functional groups (“OH and “COOH) capable of developing a negative charge and contributing substantial CEC. These large polymers possess a three-dimensional conformation that creates hydrophobic regions important in retaining nonionic synthetic organic compounds such as pesticides. Nonionic pesticides partition into these hydrophobic regions, thereby decreasing off-site movement and biological availability. A wide variety of organisms live in soil, including microorganisms not visible to the naked eye such as bacteria, fungi, protozoa, some algae, and viruses. Bacteria are present in the largest numbers, but fungi produce more biomass per unit weight of soil than any other group of microorganisms. Much of agricultural chemistry as it relates to nutrient cycles, pesticide transformation, plant growth, and organic matter degradation involves the participation of microorganisms. Microorganisms produce both intracellular and extracellular enzymes that increase reaction rates, oxidize and reduce organic and inorganic compounds, and synthesize organic molecules that modify soil chemical and physical properties.

Additional organisms in soil such as insects, nematodes, and earthworms also alter the soil ecosystem in a manner that directly or indirectly affects chemical reactions. These organisms physically process plant-derived organic materials prior to biochemical degradation by microorganisms. Nutrient release from organic materials is thus accelerated because the meso- and macrofauna expose more organic matter surface area to microbial breakdown and redistribute such materials in soil to areas of intense microbial activity.

In addition, bioturbation may also cause physical changes to the soil structure that increase pore space or modify water movement. Changes in O_2 concentration or soil water content will control biotic and abiotic reactions, altering rates of nutrient cycling and organic matter degradation.

Plant roots also modify soil by producing a zone of intense biological activity called the rhizosphere. This is a region of soil influenced by the root, most often delineated by comparing microbial numbers at greater distance from the root surface. Carbon compounds exuded or sloughed off from roots are used as a food source by microorganisms, thereby causing increased growth and activity. Microbial numbers above those of the bulk soil, which displays no root influence, indicate that the rhizosphere extends to 5 millimeters (0.2 inches) or less. Rhizosphere microorganisms that capitalize on carbon from the plant root interact physically and biochemically with the root, potentially producing positive or negative effects on plant growth.

Soil Chemistry

Biological availabilities and transport phenomena of ions and molecules in soil are controlled by the type of bonding that occurs with the solid phase. Ions such as those typically formed when amending soils with inorganic fertilizers interact with high surface area clay and humic colloids to form either outer- or inner-sphere complexes. Outer-sphere complexes result when ions, electrostatically attracted to an oppositely charged colloidal surface, retain their shell of hydrating water molecules. These loosely held ions satisfy the excess positive or negative charge of the colloid, but are separated from the colloid's surface by one or more layers of water. In contrast, inner-sphere complexes form when the ion loses its hydration water to form a much stronger covalent bond with the colloid. Nutrient ions held in outer-sphere complexes are plant-available because they may be exchanged with ions of the same charge, but nutrients held by an inner-sphere mechanism are not available until the covalent bond is broken.

Most soils contain a net CEC often reported in centimoles of charge per kilogram of soil ($cmol_c/kg$). Biological and physical characteristics of the soil are controlled by the amount of CEC and the specific cations involved. Soils dominated by high surface area clays or humus display the highest CECs, whereas soils with large amounts of sand or silt, and only small amounts of humus, exhibit much lower CECs. Highly charged cations with small hydrated radii such as Al^{3+} are more tightly held on the CEC and less likely to exchange than larger, less highly charged cations such as Na^+. This general relationship is superseded when a specific inner-sphere complex forms such as between Cu^{2+} and humus, or K^+ and clay. An even more dramatic example is that of two plant nutrients, NO_3^- and PO_4^{3-}. Negatively charged NO_3^- readily leaches out of soil, but PO_4^{3-} is retained quite strongly because it forms an inner-sphere complex.

The percentage of the CEC occupied by specific cations influences soil pH and associated characteristics relevant to plant growth and soil biological activity. Only the most strongly held cations remain in soils in high rainfall areas. Al^{3+} dominates the CEC, hydrolyzing when released from the solid phase to the soil solution to form acidic soils with pH values often below 5.

$$Al^{3+} + H_2O \leftrightarrow AlOH^{2+} + H^+$$

In contrast, soils located in lower rainfall areas accumulate less tightly bound cations such as Ca^{2+}, Mg^{2+}, K^+, and Na^+ and have higher pH values between 5 and 7. In the most arid regions, large amounts of OH^--generating sodium and calcium salts accumulate, causing soil pH values to exceed 7. Plant growth is optimal in soils having pH values between 5.5 and 6.5 because aluminum toxicity occurring at lower pH values, and nutrient limitations caused by higher pH values, are avoided.

Soil Microbiology and Biochemistry

Biochemical transformations catalyzed largely by microorganisms are required for the sustained productivity of all ecosystems. Nutrients sequestered in organic materials and added in the form of fertilizers are cycled by microorganisms in their quest for energy, reducing equivalents, and carbon. Microorganisms grow and reproduce by oxidizing organic or inorganic materials, thereby releasing electrons.

The electrons are passed down a series of carriers aligned in a thermodynamic gradient designed to capture energy in the form of adenosine triphosphate (ATP). Additional electrons originating from organic or inorganic materials are used to provide reducing equivalents necessary for synthesizing cell constituents. Carbon for cell growth is obtained from the organic materials being oxidized or captured in the form of CO_2 if inorganic materials are being oxidized.

Reduction-oxidation processes are therefore central to agricultural chemistry because oxidation of the electron source and reduction of the electron sink profoundly modify the respective element's chemical characteristics, and thus its behavior and biological availability in the environment. For example, microbial oxidation processes convert organic compounds to CO_2, a gas, and NH_4^+, a cation, to NO_3^-, an anion. Electrons obtained in these oxidations are passed on to a terminal electron acceptor. Microorganisms use terminal electron acceptors in a sequence that maximizes energy yield starting with O_2 and proceeding through NO_3^-, Mn^{4+}, Fe^{3+}, SO_4^{2-}, and finally CO_2, which upon reduction yield H_2O, N_2, Mn^{2+}, Fe^{2+}, H_2S, and CH_4, respectively.

Human Manipulation of Agricultural Ecosystems

Food and fiber production are typically optimized by carefully managing the agricultural ecosystem. Synthetic organic compounds are often applied to control plant pests including weeds, insects, nematodes, and fungal pathogens. Pesticide fate is controlled by sorption to the solid phase and degradation rate.

Because most soils have a CEC, cationic pesticides are so strongly held by soil that they are typically biologically unavailable. Weak acid pesticides containing carboxyl, phenolic hydroxyls, or aminosulfonyl functional groups are weakly retained by soil and thus most likely to leach or move off-site. Weak bases, which may exist as positively charged or uncharged molecules, and nonionic compounds, are intermediate in their susceptibility to move off-site and cause environmental contamination. However, rapid degradation of some pesticides to form benign products eliminates the time available for transport, decreasing the potential for environmental problems. Both biotic and abiotic mechanisms catalyze degradative reactions, the rate of which is controlled by the pesticide's chemical structure. With the advent of molecular techniques and the ability to transfer genes, an additional area of concern has emerged: the introduction of foreign genes into plant species for enhanced crop productivity.

In addition, we have the ability to produce a variety of pharmaceutical chemicals in genetically modified plants using what has been termed "pharm crops." David Suzuki and Holly Dressel in *From Naked Ape to Superspecies* have commented on such genetic manipulations, addressing the risks of placing genes from one species into another. Not only is direct gene transfer from one living organism to another possible, but extracellular DNA preserved in soil systems is also potentially available for transfer, further increasing environmental risks.

Agricultural chemistry is most often linked to food and fiber production, specifically for human consumption. Jared Diamond in *Guns, Germs, and Steel* argues quite convincingly that it was our ability to domesticate crops and eliminate the need for hunting and gathering that allowed for the establishment of permanent settlements and the development of technologically advanced societies.

The ensuing increase in human population has led to tremendous pressure to produce additional food from finite resources. Increased agricultural production, in combination with additional resource consumption and waste generation, has caused environmental degradation. By understanding key concepts in agricultural chemistry, we can utilize the soil resource to produce an adequate food supply and protect the environment.

Agricultural Chemistry and Soil Science

Fertilizers and manures are the crucial inputs in the modern system of coffee cultivation. Scientific use of fertilizers assumes vital importance for sustainable coffee cultivation.

The agricultural and soil science aspects of coffee cultivation forms the mandate of the faculty. Both academic and applied research in this direction are being carried out.

Trials on fertilizers, manures and micro-nutrients are being conducted. Routine soil, leaf and agro-chemical analysis for the purpose of advisory to the growers also form a part of the programme. Interdisciplinary research is the recent trend undertaken by the division.

Salient Achievements

- Optimum doses of fertilizers standardized based on the soil analysis, leaf analysis and crop removal.
- Leaf sampling techniques standardized.
- Source of nutrient specific to soil conditions determined.
- Foliar spray of nutrients in the regular fertilizer programme recommended.
- Soil fertility survey of different regions indexed with respect to both major and micro nutrients.
- Remedial measures for micro nutrient deficiency suggested.
- Role of citric acid in increasing availability of phosphorus in the soil studied.

Thrust Areas

- Development of Diagnostic Recommendation Integrated System norms for arabica coffee.
- Preparation of soil fertility map of various coffee growing zones.
- Survey and mapping of micro-nutrient status of different coffee growing tracts.
- Studies on the significance of Sulphur nutrition in coffee.

On-going Programmes

- Soil and leaf testing projects.
- Development of sex pheromone system for the integrated management of white stem borer.
- Trials on the fertilizer and proprietary formulations and agricultural chemistry aspects.

Chapter 2

Chemical Engineering

Chemical engineering is the branch of engineering that deals with the application of physical science (e.g. chemistry and physics), and life sciences (e.g. biology, microbiology and biochemistry) with mathematics, to the process of converting raw materials or chemicals into more useful or valuable forms. In addition to producing useful materials, modern chemical engineering is also concerned with pioneering valuable new materials and techniques-such as nanotechnology, fuel cells and biomedical engineering. A person employed in this field is called a chemical engineer.

Chemical engineering largely involves the design, improvement and maintenance of processes involving chemical or biological transformations for large-scale manufacture. Chemical engineers ensure the processes are operated safely, sustainably and economically. Chemical engineers in this branch are usually employed under the title of process engineer. A related term with a wider definition is chemical technology.

In 1824, French physicist Sadi Carnot, in his "On the Motive Power of Fire", was the first to study the thermodynamics of combustion reactions in steam engines. In the 1850s, German physicist Rudolf Clausius began to apply the principles developed by Carnot to chemical systems at the atomic to molecular scale.

During the years 1873 to 1876 at Yale University, American mathematical physicist Josiah Willard Gibbs, the first to be awarded a Ph.D. in engineering in the U.S., in a series of three papers, developed a mathematical-based, graphical methodology, for the study of chemical systems using the thermodynamics of Clausius. In 1882, German physicist Hermann von Helmholtz, published a founding thermodynamics paper, similar to Gibbs, but with more of an electro-chemical basis, in which he showed that measure of chemical affinity, i.e. the "force" of chemical reactions, is determined by the measure of the free energy of the reaction process. Following these early developments, the new science of chemical engineering began to develop. The following timeline shows some of the key steps in the development of the science of chemical engineering:

- 1805 – John Dalton published Atomic Weights, allowing chemical equations to be balanced and the basis for chemical engineering mass balances.
- 1882 – a course in "Chemical Technology" is offered at University College London
- 1883 – Osborne Reynolds defines the dimensionless group for fluid flow, leading to practical scale-up and understanding of flow, heat and mass transfer
- 1885 – Henry Edward Armstrong offers a course in "chemical engineering" at Central College (later Imperial College), London.
- 1888 – There is a Department of Chemical Engineering at Glasgow and West of Scotland Technical College offering day and evening classes.
- 1888 – Lewis M. Norton starts a new curriculum at Massachusetts Institute of Technology (MIT): Course X, Chemical Engineering
- 1889 – Rose Polytechnic Institute awards the first bachelor's of science in chemical engineering in the US.
- 1891 – MIT awards a bachelor's of science in chemical engineering to William Page Bryant and six other candidates.
- 1892 – A bachelor's program in chemical engineering is established at the University of Pennsylvania.
- 1901 – George E. Davis produces the *Handbook of Chemical Engineering*
- 1905 – the University of Wisconsin awards the first Ph.D. in chemical engineering to Oliver Patterson Watts.
- 1908 – the American Institute of Chemical Engineers (AIChE) is founded.
- 1922 – the UK Institution of Chemical Engineers (IChemE) is founded.

- 1942 – Hilda Derrick, first female student member of the IChemE.

Applications

Chemical engineering is applied in the manufacture of a wide variety of products. The chemical industry proper manufactures inorganic and organic industrial chemicals, ceramics, fuels and petrochemicals, agrochemicals (fertilizers, insecticides, herbicides), plastics and elastomers, oleochemicals, explosives, detergents and detergent products (soap, shampoo, cleaning fluids), fragrances and flavors, additives, dietary supplements and pharmaceuticals. Closely allied or overlapping disciplines include wood processing, food processing, environmental technology, and the engineering of petroleum, glass, paints and other coatings, inks, sealants and adhesives.

Overview

Chemical engineers design processes to ensure the most economical operation. This means that the entire production chain must be planned and controlled for costs. A chemical engineer can both simplify and complicate "showcase" reactions for an economic advantage. Using a higher pressure or temperature makes several reactions easier; ammonia, for example, is simply produced from its component elements in a high-pressure reactor. On the other hand, reactions with a low yield can be recycled continuously, which would be complex, arduous work if done by hand in the laboratory. It is not unusual to build 6-step, or even 12-step evaporators to reuse the vaporization energy for an economic advantage. In contrast, laboratory chemists evaporate samples in a single step.

The individual processes used by chemical engineers (eg. distillation or filtration) are called unit operations and consist of chemical reactions, mass-, heat- and momentum- transfer operations. Unit operations are grouped together in various configurations for the purpose of chemical synthesis and/or chemical separation. Some processes are a combination of intertwined transport and separation unit operations, (e.g. reactive distillation).

Three primary physical laws underlying chemical engineering design are conservation of mass, conservation of momentum and conservation of energy. The movement of mass and energy around a chemical process are evaluated using mass balances and energy balances, laws that apply to discrete parts of equipment, unit operations, or an entire plant. In doing so, chemical engineers must also use principles of thermodynamics, reaction kinetics and transport phenomena. The task of performing these balances is now aided by process simulators, which are complex software models that can solve mass and energy balances and usually have built-in modules to simulate a variety of common unit operations.

Modern Chemical Engineering

The modern discipline of chemical engineering encompasses much more than just process engineering. Chemical engineers are now engaged in the development and production of a diverse range of products, as well as in commodity and specialty chemicals. These products include high performance materials needed for aerospace, automotive, biomedical, electronic, environmental, space and military applications. Examples include ultra-strong fibres, fabrics, dye-sensitized solar cells, adhesives and composites for vehicles, bio-compatible materials for implants and prosthetics, gels for medical applications, pharmaceuticals, and films with special dielectric, optical or spectroscopic properties for opto-electronic devices.

Additionally, chemical engineering is often intertwined with biology and biomedical engineering. Many chemical engineers work on biological projects such as understanding biopolymers (proteins) and mapping the human genome. The line between chemists and chemical engineers is growing ever more thin as more and more chemical engineers begin to start their own innovation using their knowledge of chemistry, physics and mathematics to create, implement and mass produce their ideas.

History of Chemical Engineering

Chemical engineering as a discipline is a little over one hundred years old. It grew out of mechanical engineering in the last part of the 19th century, because of a need for chemical processors. Before the Industrial Revolution (18th century), industrial chemicals were mainly produced through batch processing. Batch processing is similar to cooking. Individuals would mix ingredients into a vessel, heat or pressurize the mixture, test it, and purify it to get a salable product. Batch processes are still performed today on expensive products, such as perfumes, or pure maple syrups, where one can still turn a profit, despite batch methods being slow and inefficient. Most chemicals today are produced through a continuous "assembly line" chemical process. The Industrial Revolution was when this shift from batch to continuous processing occurred.

Origin

The Industrial Revolution led to an unprecedented escalation in demand, both with regard to quantity and quality, for bulk chemicals such as sulfuric acid and soda ash. This meant two things: one, the size of the activity and the efficiency of operation had to be enlarged, and two, serious alternatives to batch processing, such as continuous operation, had to be examined.

This created the need for an engineer who was not only conversant with how machines behaved, but also understood chemical reactions and transport phenomena (how substances came together to react, how the required conditions could be achieved, etc.), and the influence the equipment had on how these processes operated on the large scale. Thus, Chemical Engineering was born as a distinct discipline; distinct from both Mechanical Engineering on one hand and industrial chemistry on the other.

Professional Associations

These early programs married industrial chemistry with mechanical engineering, with the emphasis most decidedly on engineering. But chemical engineers still needed to clearly define their activity as something more than a mishmash of chemistry and engineering. To emphasize their identity and thus help the growth of their profession, chemical engineers formed the American Institute of Chemical Engineers in 1908. The Institution of Chemical Engineers was founded in 1922 and awarded a Royal Charter in 1957.

Definitions

For the other established branches of engineering, there were ready associations in the mind of the common man: Mechanical Engineering meant machines, Electrical Engineering meant circuitry, and Civil Engineering meant structures. So what symbolized chemical engineering?

Unit Operation

The answer, provided by Arthur D. Little to the President of MIT, was to emphasize the approach chemical engineers took to the design and analysis of processes rather than a process or a product. The concept of Unit operations was developed to emphasize the underlying unity among seemingly different operations.

For example, the principles are the same whether one is concerned about separating alcohol from water in a fermenter, or separating gasoline from diesel in a refinery, as long as the basis of separation is generation of a vapour of a different composition from the liquid. Therefore such separation processes can be studied together as a unit operation (in this case called distillation). The concept has stood the profession in good stead in its phase of growth, and has even been used to understand the way the human body functions.

Unit Processes

In the early part of the last century, a parallel concept called Unit Processes was used to classify reactive processes. Thus oxidations, reductions, alkylations etc. formed separate unit processes and were studied as such. This was natural considering the close affinity of chemical engineering to industrial chemistry at its inception. Gradually however, the subject of chemical reaction engineering has largely replaced the unit process concept. This subject looks at the entire body of chemical reactions as having a personality of its own, independent of the particular chemical species or chemical bonds involved. The latter does contribute to this personality in no small measure, but to design and operate chemical reactors, a knowledge of characteristics such as rate behaviour, thermodynamics, single or multiphase nature, etc. are more important. The emergence of chemical reaction engineering as a discipline truly signalled the severance of the umbilical cord connecting chemical engineering to industrial chemistry, and served to cement the truly unique character of this discipline.

Chemical Bond

A chemical bond is the physical process responsible for the attractive interactions between atoms and molecules, and that which confers stability to diatomic and polyatomic chemical compounds. The explanation of the attractive forces is a complex area that is described by the laws of quantum electrodynamics. In practice, however, chemists usually rely on quantum theory or qualitative descriptions that are less rigorous but more easily explained to describe chemical bonding. In general, strong chemical bonding is associated with the sharing or transfer of electrons between the participating atoms. Molecules, crystals, and diatomic gases—indeed most of the physical environment around us—are held together by chemical bonds, which dictate the structure of matter.

Bonds vary widely in their strength which is associated both with the energy required to break them, and the forces they exert on the atoms they hold together. Generally covalent and ionic bonds are often described as "strong", whereas hydrogen bonds and van der Waals' bonds are generally considered to be "weak," although there exist overlaps in strength within these bond classes.

Since opposite charges attract via a basic electromagnetic force, the negatively-charged electrons orbiting the nucleus and the positively-charged protons in the nucleus attract each other. Also, an electron positioned between two nuclei will be attracted to both of them. Thus, the most stable configuration of nuclei and electrons is one in which the electrons spend more time *between* nuclei, than anywhere else in space. These electrons cause the nuclei to be attracted to each other. However, this assembly cannot collapse to a size dictated by the volumes of these individual particles. Due to the matter wave nature of electrons and their smaller mass, they occupy a very much larger amount of volume compared with the nuclei, and this volume occupied by the electrons keeps the atomic nuclei relatively far apart, as compared with the size of the nuclei themselves.

In the simplest view of a so-called *covalent* bond, one or more electrons (often a pair of electrons) are drawn into the space between the two atomic nuclei. Here the negatively charged electrons are attracted to the positive charges of *both* nuclei, instead of just their own. This overcomes the repulsion between the two positively charged nuclei of the two atoms, and so this overwhelming attraction holds the two nuclei in a fixed configuration of equilibrium, even though they will still vibrate at equilibrium position. In summary, covalent bonding involves sharing of electrons in which the positively charged nuclei of two or more atoms simultaneously attract the negatively charged electrons that are being shared. In a polar covalent bond, one or more electrons are unequally shared between two nuclei.

In a simplified view of an *ionic* bond, the bonding electron is not shared at all, but transferred. In this type of bond, the outer atomic orbital of one atom has a vacancy which allows addition of one or more electrons. These newly added electrons potentially occupy a lower energy-state (effectively closer to more nuclear charge) than they experience in a different atom. Thus, one nucleus offers a more tightly-bound position to an electron than does another nucleus, with the result that one atom may transfer an electron to the other. This transfer causes one atom to assume a net positive charge, and the other to assume a net negative charge. The *bond* then results from electrostatic attraction between atoms, and the atoms become positive or negatively charged ions.

All bonds can be explained by quantum theory, but, in practice, simplification rules allow chemists to predict the strength, directionality, and polarity of bonds. The octet rule and VSEPR theory are two examples. More sophisticated theories are valence bond theory which includes orbital hybridization and resonance, and the linear combination of atomic orbitals molecular orbital method which includes ligand field theory. Electrostatics are used to describe bond polarities and the effects they have on chemical substances.

History

Early speculations into the nature of the chemical bond, from as early as the 12th century, supposed that certain types of chemical species were joined by a type of chemical affinity. In 1704, Isaac Newton famously outlined his atomic bonding theory, in "Query 31" of his Opticks, whereby atoms attach to each other by some "force". Specifically, after acknowledging the various popular theories in vogue at the time, of how atoms were reasoned to attach to each other, i.e. "hooked atoms", "glued together by rest", or "stuck together by conspiring motions", Newton states that he would rather infer from their cohesion, that "particles attract one another by some force, which in immediate contact is exceedingly strong, at small distances performs the chemical operations, and reaches not far from the particles with any sensible effect."

In 1819, on the heels of the invention of the voltaic pile, Jöns Jakob Berzelius developed a theory of chemical combination stressing the electronegative and electropositive character of the combining atoms. By the mid 19th century, Edward Frankland, F.A. Kekule, A.S. Couper, A.M. Butlerov, and Hermann Kolbe, building on the theory of radicals, developed the theory of valency, originally called "combining power", in which compounds were joined owing to an attraction of positive and negative poles. In 1916, chemist Gilbert N. Lewis developed the concept of the electron-pair bond, in which two atoms may share one to six electrons, thus forming the single electron bond, a single bond, a double bond, or a triple bond; in Lewis's own words, "An electron may form a part of the shell of two different atoms and cannot be said to belong to either one exclusively."

That same year, Walther Kossel put forward a theory similar to Lewis' only his model assumed complete transfers of electrons between atoms, and was thus a model of ionic bonds. Both Lewis and Kossel structured their bonding models on that of Abegg's rule (1904).

In 1927, the first mathematically complete quantum description of a simple chemical bond, i.e. that produced by one electron in the hydrogen molecular ion, H_2^+, was derived by the Danish physicist Oyvind Burrau. This work showed that the quantum approach to chemical bonds could be fundamentally and quantitatively correct, but the mathematical methods used could not be extended to molecules containing more than one electron. A more practical, albeit less quantitative, approach was put forward in the same year by Walter Heitler and Fritz London. The Heitler-London method forms the basis of what is now called valence bond theory. In 1929, the linear combination of atomic orbitals molecular orbital method (LCAO) approximation was introduced by Sir John Lennard-Jones, who also suggested methods to derive electronic structures of molecules of F_2 (fluorine) and O_2 (oxygen) molecules, from basic quantum principles. This molecular orbital theory represented a covalent bond as a orbitals formed by combining the quantum mechanical Schrödinger atomic orbitals which had been hypothesized for electrons in single atoms. The equations for bonding electrons in multi-electron atoms could not be solved to mathematical perfection (i.e., *analytically*), but approximations for them still gave many good qualitative predictions and results. Most quantitative calculations in modern quantum chemistry use either valence bond or molecular orbital theory as a starting point, although a third approach, Density Functional Theory, has become increasingly popular in recent years.

In 1935, H. H. James and A. S. Coolidge carried out a calculation on the dihydrogen molecule that, unlike all previous calculation which used functions only of the distance of the electron from the atomic nucleus, used functions which also explicitly added the distance between the two electrons. With up to 13 adjustable parameters they obtained a result very close to the experimental result for the dissociation energy. Later extensions have used up to 54 parameters and give excellent agreement with experiment. This calculation convinced the scientific community that quantum theory could give agreement with experiment. However this approach has none of the physical pictures of the valence bond and molecular orbital theories and is difficult to extend to larger molecules.

Valence Bond Theory

In the year 1927, valence bond theory was formulated which argued essentially that a chemical bond forms when two valence electrons, in their respective atomic orbitals, work or function to hold two nuclei together, by virtue of system energy lowering effects. In 1931, building on this theory, chemist Linus Pauling published what some consider one of the most important papers in the history of chemistry: "On the Nature of the Chemical Bond". In this paper, building on the works of Lewis, and the valence bond theory (VB) of Heitler and London, and his own earlier work, he presented six rules for the shared electron bond, the first three of which were already generally known:

1. The electron-pair bond forms through the interaction of an unpaired electron on each of two atoms.
2. The spins of the electrons have to be opposed.
3. Once paired, the two electrons cannot take part in additional bonds. His last three rules were new:
4. The electron-exchange terms for the bond involves only one wave function from each atom.
5. The available electrons in the lowest energy level form the strongest bonds.
6. Of two orbitals in an atom, the one that can overlap the most with an orbital from another atom will form the strongest bond, and this bond will tend to lie in the direction of the concentrated orbital.

Building on this article, Pauling's 1939 textbook: *On the Nature of the Chemical Bond* would become what some have called the "bible" of modern chemistry. This book helped experimental chemists to understand the impact of quantum theory on chemistry. However, the later edition in 1959 failed to address adequately the problems that appeared to be better understood by molecular orbital theory. The impact of valence theory declined during the 1960's and 1970's as molecular orbital theory grew in popularity and was implemented in many large computer programs. Since the 1980s, the more difficult problems of implementing valence bond theory into computer programs have been largely solved and valence bond theory has seen a resurgence.

Comparison of Valence Bond and Molecular Orbital Theory

In some respects valence bond theory is superior to molecular orbital theory. When applied to the simplest two-electron molecule, H_2, valence bond theory, even at the simplest Heitler-London approach, gives a much closer approximation to the bond energy, and it provides a much more accurate representation of the behavior of the electrons as chemical bonds are formed and broken.

In contrast simple molecular orbital theory predicts that the hydrogen molecule dissociates into a linear superposition of hydrogen atoms and positive and negative hydrogen ions, a completely unphysical result. This explains in part why the curve of total energy against interatomic distance for the valence bond method lies above the curve for the molecular orbital method at all distances and most particularly so for large distances. This situation arises for all homonuclear diatomic molecules and is particularly a problem for F_2, where the minimum energy of the curve with molecular orbital theory is still higher in energy than the energy of two F atoms.

The concepts of hybridization are so versatile, and the variability in bonding in most organic compounds is so modest, that valence bond theory remains an integral part of the vocabulary of organic chemistry. However, the work of Friedrich Hund, Robert Mulliken, and Gerhard Herzberg showed that molecular orbital theory provided a more appropriate description of the spectroscopic, ionization and magnetic properties of molecules. The deficiencies of valence bond theory became apparent when hypervalent molecules (e.g. PF_5) were explained without the use of d orbitals that were crucial to the bonding hybridisation scheme proposed for such molecules by Pauling. Metal complexes and electron deficient compounds (e.g. diborane) also appeared to be well described by molecular orbital theory, although valence bond descriptions have been made.

In the 1930s the two methods strongly competed until it was realised that they are both approximations to a better theory. If we take the simple valence bond structure and mix in all possible covalent and ionic structures arising from a particular set of atomic orbitals, we reach what is called the full configuration interaction wave function.

If we take the simple molecular orbital description of the ground state and combine that function with the functions describing all possible excited states using unoccupied orbitals arising from the same set of atomic orbitals, we also reach the full configuration interaction wavefunction. It can be then seen that the simple molecular orbital approach gives too much weight to the ionic structures, while the simple valence bond approach gives too little. This can also be described as saying that the molecular orbital approach is too *delocalised*, while the valence bond approach is too *localised*.

The two approaches are now regarded as complementary, each providing its own insights into the problem of chemical bonding. Modern calculations in quantum chemistry usually start from (but ultimately go far beyond) a molecular orbital rather than a valence bond approach, not because of any intrinsic superiority in the former but rather because the MO approach is more readily adapted to numerical computations. However better valence bond programs are now available.

Bonds in Chemical Formula

The 3-dimensionality of atoms and molecules makes it difficult to use a single technique for indicating orbitals and bonds. In molecular formulae the chemical bonds (binding orbitals) between atoms are indicated by various different methods according to the type of discussion. Sometimes, they are completely neglected.

For example, in organic chemistry chemists are sometimes concerned only with the functional groups of the molecule. Thus, the molecular formula of ethanol (a compound in alcoholic beverages) may be written in a paper in conformational, 3-dimensional, full 2-dimensional (indicating every bond with no 3-dimensional directions), compressed 2-dimensional (CH_3–CH_2–OH), separating the functional group from another part of the molecule (C_2H_5OH), or by its atomic constituents (C_2H_6O), according to what is discussed.

Sometimes, even the non-bonding valence shell electrons (with the 2-dimensional approximate directions) are marked, i.e. for elemental carbon.'C'.

Strong Chemical Bonds

Typical bond lengths in pmand bond energies in kJ/mol.Bond lengths can be converted to Åby division by 100 (1 Å = 100 pm).Data taken from.

Bond	Length(pm)	Energy(kJ/mol)
H — Hydrogen		
H–H	74	436
H–C	109	413
H–N	101	391
H–O	96	366
H–F	92	568
H–Cl	127	432
H–Br	141	366
N — Nitrogen		
N–H	101	391
N–C	147	308
N–N	145	170
Na"N	110	945
C — Carbon		
C–H	109	413
C–C	154	348
C=C	134	614
Ca"C	120	839
C–N	147	308
C–O	143	360
C–F	134	488
C–Cl	177	330
C–Br	194	288
C–I	214	216
C–S	182	272

O — Oxygen		
O–H	96	366
O–C	143	360
O–O	148	145
O=O	121	498
F, Cl, Br, I — Halogens		
F–H	92	568
F–F	142	158
F–C	135	488
Cl–H	127	432
Cl–C	177	330
Cl–Cl	199	243
Br–H	141	366
Br–C	194	288
Br–Br	228	193
I–H	161	298
I–C	214	216
I–I	267	151
S — Sulfur		
C–S	182	272

These chemical bonds are *intramolecular* forces, which hold atoms together in molecules. In the simplistic localized view of bonding, the number of electrons participating in a bond (or located in a bonding orbital) is typically multiples of two, four, or six, respectively. Even numbers are common because electrons enjoy lower energy states, if paired. Substantially more advanced bonding theories have shown that bond strength is not always a whole number, depending on the distribution of electrons to each atom involved in a bond. For example, the carbons in benzene are connected to each other with about 1.5 bonds, and the two atoms in nitric oxide NO, are connected with about 2.5 bonds. Quadruple bonds are also well known. The type of strong bond depends on the difference in electronegativity and the distribution of the electron orbital paths available to the atoms that are bonded. The larger the difference in electronegativity, the more an electron is attracted to a particular atom involved in the bond, and the more "ionic" properties the bond is said to have ("ionic" means the bond electron(s) are unequally shared). The smaller the difference in electronegativity, the more covalent properties (full sharing) the bond has.

Ionic Bond

Ionic bonding is a type of electrostatic interaction between atoms which have a large electronegativity difference. There is no precise value that distinguishes ionic from covalent bonding but a difference of electronegativity of over 1.7 is likely to be ionic and a difference of less than 1.7 is likely to be covalent. Ionic bonding leads to separate positive and negative ions. Ionic charges are commonly between "3e to +3e.

Polar Covalent Bond

Polar covalent bonding is intermediate in character between a covalent and an ionic bond.

Coordinate Covalent Bond

Coordinate covalent bonding, sometimes referred to as dative bonding, is a kind of covalent bonding, in which the covalent bonding electrons originate solely from one of the atoms, the electron-pair donor or Lewis base but are approximately equally shared in the formation of a covalent bond.

This concept is somewhat fading as chemists increasingly embrace molecular orbital theory. Examples of coordinate covalent bonding occur in nitrones and ammonia borane. The arrangement is different from an ionic bond in that the electronegativity difference is small, resulting in covalency. They are shown by an arrow unlike other bonds. This arrow shows its head towards the electron acceptor or lewis acid and its tail towards the lewis base. This bond is seen in ammonium.

One- and Three-electron Bonds

Bonds with one or three electrons can be found in radical species, which have an odd number of electrons. The simplest example of a 1-electron bond is found in the hydrogen molecular cation, H_2^+. One-electron bonds often have about half the bond energy of a 2-electron bond, and are therefore called "half bonds". However, there are exceptions: in the case of dilithium, the bond is actually stronger for the 1-electron Li_2^+ than for the 2-electron Li_2. This exception can be explained in terms of hybridization and inner-shell effects.

The simplest example of three-electron bonding can be found in the helium dimer cation, He_2^+, and can also be considered a "half bond" because, in molecular orbital terms, the third electron is in an anti-bonding orbital which cancels out half of the bond formed by the other two electrons. Another example of a molecule containing a 3-electron bond, in addition to two 2-electron bonds, is nitric oxide, NO. The oxygen molecule, O_2 can also be regarded as having two 3-electron bonds and one 2-electron bond, which accounts for its paramagnetism and its formal bond order of 2.

Molecules with odd-electron bonds are usually highly reactive. These types of bond are only stable between atoms with similar electronegativities.

Bent Bonds

Bent bonds, also known as banana bonds, are bonds in strained or otherwise sterically hindered molecules whose binding orbitals are forced into a banana-like form. Bent bonds are often more susceptible to reactions than ordinary bonds.

3c-2e and 3c-4e Bonds

In three-centre two-electron bonds ("3c-2e") three atoms share two electrons in bonding. This type of bonding occurs in electron deficient compounds like diborane. Each such bond (2 per molecule in diborane) contains a pair of electrons which connect the boron atoms to each other in a banana shape, with a proton (nucleus of a hydrogen atom) in the middle of the bond, sharing electrons with both boron atoms. Three-centre four-electron bonds ("3c-4e") also exist which explain the bonding in hypervalent molecules. In certain cluster compounds, so-called four-centre two-electron bonds also have been postulated. In certain conjugated π (pi) systems, such as benzene and other aromatic compounds, and in conjugated network solids such as graphite, the electrons in the conjugated system of π-bonds are spread over as many nuclear centers as exist in the molecule or the network.

Aromatic Bond

In most cases, the locations of electrons cannot be simplified to simple lines or dots (a single electron). In aromatic bonds which occur in planar rings of atoms where the 4n+2 rule determines whether ring molecules would show extra stability.

In benzene, the prototypical aromatic compound, 18 bonding electrons bind 6 carbon atoms together to form a planar ring structure. The bond "order" (average number of bonds) between the different carbon atoms may be said to be (18/6)/2=1.5, but in this case the bonds are all identical from the chemical point of view. They may sometimes be written as single bonds alternating with double bonds, but the view of all ring bonds as being equivalently about 1.5 bonds in strength, is much closer to truth.

In the case of heterocyclic aromatics and substituted benzenes, the electronegativity differences between different parts of the ring may dominate the chemical behaviour of aromatic ring bonds, which otherwise are equivalent.

Metallic Bond

In a metallic bond, bonding electrons are delocalized over a lattice of atoms. By contrast, in ionic compounds, the locations of the binding electrons and their charges are static. Because of delocalization or the free moving of electrons, it leads to the metallic properties such as conductivity, ductility and hardness.

Intermolecular Bonding

There are four basic types of bonds that can be formed between two or more (otherwise non-associated) molecules, ions or atoms. Intermolecular forces cause molecules to be attracted or repulsed by each other. Often, these define some of the physical characteristics (such as the melting point) of a substance.

Permanent Dipole to Permanent Dipole

A large electronegativity difference between two strongly bonded atoms within a molecule causes a dipole to form (a dipole is a pair of permanent partial charges). Dipoles will attract or repel each other.

Hydrogen Bond

In some ways this is an especially strong example of a permanent dipole, as above. However, in the hydrogen bond, the hydrogen proton comes closer to being shared between target and donor atoms, in a three-centre two-electron bond like that in diborane. Hydrogen bonds explain the relatively high boiling points of liquids like water, ammonia, and hydrogen fluoride, compared with their heavier counterparts in the same periodic table column.

Instantaneous Dipole to Induced Dipole (Van Der Waals)

Instantaneous dipole to induced dipole, or van der Waals forces, are the weakest, but also the most prolific—occurring between all chemical substances. Imagine a helium atom: At any one point in time, the electron cloud around the (otherwise neutral) atom can be thought to be slightly imbalanced, with momentarily more negative charge on one side. This is referred to as an instantaneous dipole. This dipole, with its slight charge imbalance, may attract or repel the electrons within a neighbouring helium atom, setting up another dipole. The two atoms will be attracted for an instant, before the charge rebalances and the atoms move on.

Cation-pi Interaction

Cation-pi interactions occur between the localized negative charge of π orbital electrons, located above and below the plane of an aromatic ring, and a positive charge.

Electrons in Chemical Bonds

In the (unrealistic) limit of "pure" ionic bonding, electrons are perfectly localized on one of the two atoms in the bond. Such bonds can be understood by classical physics. The forces between the atoms are characterized by isotropic continuum electrostatic potentials. Their magnitude is in simple proportion to the charge difference. Covalent bonds are better understood by valence bond theory or molecular orbital theory. The properties of the atoms involved can be understood using concepts such as oxidation number. The electron density within a bond is not assigned to individual atoms, but is instead delocalized between atoms. In valence bond theory, the two electrons on the two atoms are coupled together with the bond strength depending on the overlap between them. In molecular orbital theory, the linear combination of atomic orbitals (LCAO) helps describe the delocalized molecular orbital structures and energies based on the atomic orbitals of the atoms they came from. Unlike pure ionic bonds, covalent bonds may have directed anisotropic properties. These may have their own names, such as Sigma and Pi bond.

In the general case, atoms form bonds that are intermediates between ionic and covalent, depending on the relative electronegativity of the atoms involved. This type of bond is sometimes called polar covalent.

Covalent Bond

A covalent bond is a form of chemical bonding that is characterized by the *sharing* of pairs of electrons between atoms, or between atoms and other covalent bonds. In short, attraction-to-repulsion stability that forms between atoms when they share electrons is known as covalent bonding.

Covalent bonding includes many kinds of interaction, including σ-bonding, π-bonding, metal to non-metal bonding, agostic interactions, and three-centre two-electron bonds. The term *covalent bond* dates from 1939. The prefix *co-* means *jointly, associated in action, partnered to a lesser degree,* etc.; thus a "co-valent bond", essentially, means that the atoms share "valence", such as is discussed in valence bond theory. In the molecule H_2, the hydrogen atoms share the two electrons via covalent bonding. Covalency is greatest between atoms of similar electronegativities. Thus, covalent bonding does not necessarily require the two atoms be of the same elements, only that they be of comparable electronegativity. Although covalent bonding entails sharing of electrons, it is not necessarily delocalized. Furthermore, in contrast to electrostatic interactions ("ionic bonds") the strength of covalent bond depends on the angular relation between atoms in polyatomic molecules.

History

The term "covalence" in regard to bonding was first used in 1919 by Irving Langmuir in a *Journal of the American Chemical Society* article entitled "The Arrangement of Electrons in Atoms and Molecules". Langmuir wrote that "we shall denote by the term covalence the number of pairs of electrons which a given atom shares with its neighbors." The idea of covalent bonding can be traced several years before 1919 to Gilbert N. Lewis, who in 1916 described the sharing of electron pairs between atoms. He introduced the so called *Lewis notation* or *electron dot notation or The Lewis Dot Structure* in which valence electrons (those in the outer shell) are represented as dots around the atomic symbols. Pairs of electrons located between atoms represent covalent bonds. Multiple pairs represent multiple bonds, such as double and triple bonds. An alternative form of representation, not shown here, has bond-forming electron pairs represented as solid lines.

While the idea of shared electron pairs provides an effective qualitative picture of covalent bonding, quantum mechanics is needed to understand the nature of these bonds and predict the structures and properties of simple molecules. Walter Heitler and Fritz London are credited with the first successful quantum mechanical explanation of a chemical bond, specifically that of molecular hydrogen, in 1927. Their work was based on the valence bond model, which assumes that a chemical bond is formed when there is good overlap between the atomic orbitals of participating atoms. These atomic orbitals are known to have specific angular relationships between each other, and thus the valence bond model can successfully predict the bond angles observed in simple molecules.

Bond Order

Bond order is a number that indicates the number of pairs of electrons shared between atoms forming a covalent bond. The term is only applicable to diatomic molecules, but is used to describe bonds within polyatomic compounds as well.

1. The most common type of covalent bond is the single bond, the sharing of only one pair of electrons between two atoms. It usually consists of one sigma bond. All bonds with more than one shared pair are called multiple bonds.
2. Sharing two pairs is called a double bond. An example is in ethylene (between the carbon atoms). It usually consists of one sigma bond and one pi bond.
3. Sharing three pairs is called a triple bond. An example is in hydrogen cyanide (between C and N). It usually consists of one sigma bond and two pi-bonds.
4. Quadruple bonds are found in the transition metals. Molybdenum and rhenium are the elements most commonly observed with this bonding configuration. An example of a quadruple bond is also found in Di-tungsten tetra(hpp).
5. Quintuple bonds have been found to exist in certain dichromium compounds.
6. Sextuple bonds are found in diatomic molybdenum and tungsten.

Most bonding of course, is not localized, so the above classification, while powerful and pervasive, is of limited validity. Three-centre bonds do not conform readily to the above conventions.

Resonance

Many bonding situations can be described with more than one valid Lewis Dot Structure (for example, ozone, O_3). In an LDS diagram of O_3, the centre atom will have a single bond with one atom and a double bond with the other. The LDS diagram cannot tell us which atom has the double bond; the first and second adjoining atoms have equal chances of having the double bond. These two possible structures are called resonance structures. In reality, the structure of ozone is a resonance hybrid between its two possible resonance structures. Instead of having one double bond and one single bond, there are actually two 1.5 bonds with approximately three electrons in each at all times.

A special resonance case is exhibited in aromatic rings of atoms (for example, benzene). Aromatic rings are composed of atoms arranged in a circle (held together by covalent bonds) that may alternate between single and double bonds according to their LDS. In actuality, the electrons tend to be disambiguously and evenly spaced within the ring. Electron sharing in aromatic structures is often represented with a ring inside the circle of atoms. Lewis Dot Structures for molecules with resonance are shown by creating the dot structure for every possible form, placing brackets around each structure, and connecting the boxes with double-headed arrows.

Current Theory

Today the valence bond model has been augmented by the molecular orbital model. In this model, as atoms are brought together, the *atomic* orbitals interact to form *molecular* orbitals, which are linear sums and differences of the atomic orbitals. These molecular orbitals are a cross between the original atomic orbitals and generally extend between the two bonding atoms.

Using quantum mechanics, it is possible to calculate the electronic structure, energy levels, bond angles, bond distances, dipole moments, and electromagnetic spectra of simple molecules with a high degree of accuracy. Bond distances and angles can be calculated as accurately as they can be measured (distances to a few pm and bond angles to a few degrees). For small molecules, calculations are sufficiently accurate to be useful for determining thermodynamic heats of formation and kinetic activation energy barriers.

Types of Chemical Bounding

Atoms are the building blocks of all substances. But what is it that keeps atoms connected together? They are held together by Chemical Bonds, strong attractive forces between atoms. Without these ties that bind, the universe would be nothing more than a mass chaos of individual atoms.

So what constitutes a chemical bond? A bond is formed when electrons from two atoms interact with each other and their atoms become joined. The electrons that interact with each other are Valence Electrons, the ones that reside in the outermost electron shell of an atom.

As you sit reading this sentence, billions of atoms surround you. When any two of them approach each other, their outermost electrons come into close contact. Recall that electrons are negatively charged particles, so a chance meeting between atoms usually results in the two repelling each other. Occasionally, however, atoms will stick together by forming a chemical bond.

There are two main types of bonding discussed here. A COVALENT BOND results when two atoms "share" valence electrons between them. An IONIC BOND occurs when one atom gains a valence electron from a different atom, forming a negative ion (Anion) and a positive ion (Cation), respectively. These oppositely charged ions are attracted to each other, forming an ionic bond.

Why are chemical bonds important? The type of chemical bond that occurs in a molecule or substance in part defines its properties. For example, consider sodium chloride (NaCl) and hydrogen chloride(HCl). Both substances contain chlorine, but NaCl is the white solid crystalline substance sprinkled on French fries, and HCl is a foul smelling gas.(note: when this gas is dissolved in water, it forms a solution known as hydrochloric acid. This is the acid that your stomach uses to digest food.)

How can this be if both materials have chlorine in them? The chemical bonding that takes place in NaCl is different than that in HCl. This gives NaCl and HCl very different structures, appearances, and properties. What other differences are noticeable among molecules that result from different types of chemical bonding? Think about what happens when a bunch of sodium and chlorine ions join together to form rock salt. If we hit this with a hammer, it shatters into tiny pieces. It does this because the bonds between the atoms in rock salt are ionic. The particles are arranged in such a way that they line up along rows of positive and negative charge. Under enough stress, the salt crystals break along those lines into much smaller pieces.

The bonds that hold the carbon and hydrogen atoms in rubber together, on the other hand, are not ionic but covalent. Each carbon atom shares four of its outermost electrons with its immediate neighbors. Under stress, the bonds stretch, then snap back as each atom pulls on the shared electrons. And that's the way the ball bounces.

There is a third type of bonding, called Metallic Bonding. As the name implies, metallic bonding usually occurs in metals, such as copper. A piece of copper metal has a certain arrangement of copper atoms. The valence electrons of these atoms are free to move about the piece of metal and are attracted to the positive cores of copper, thus holding the atoms together.

Essential to understanding all types of chemical bonding is realizing that all bonds use electron "glue." Every substance is made up of atoms, and all atoms are surrounded by the charged particles called electrons. Electrons in the outermost shell can and do bond with those of other atoms to create everything from quartz and quarters to plastic and pennies. In large part, the difference between materials as diverse as diamonds and pencils is how they're glued together.

Atoms that Share

Don't look now, but you're surrounded by covalent bonds. They're in the air you breathe and in the water you drink. You even make them yourself when you exhale. What are these bonds? What do they want?

Take a look at water (H_2O). H_2O is a Molecule, a discrete unit of atoms that are bonded together. A molecule of water consists of an oxygen atom that is bonded to two hydrogen atoms. A hydrogen atom and the oxygen atom each donate one valence electron to form a chemical bond. These two valence electrons forming the bond are shared by both atoms, resulting in a Single Covalent Bond. Think of this in terms of two pieces of wood that are nailed together. The pieces of wood are the atoms, and the nails holding them together are the electrons that form the covalent bond. Each piece of wood shares a portion of the nails.

Air, too, contains oxygen. Oxygen does not exist as a single oxygen atom, but as a molecule of two oxygen atoms. These two oxygen atoms share two pairs of valence electrons (four valence electrons total) between them, forming a Double Covalent Bond. This is true of any double covalent bond; four valence electrons are shared between two atoms.

Another component of air is nitrogen. Like oxygen, nitrogen does not exist as a single nitrogen atom, but as a molecule made up of two nitrogen atoms. The two nitrogen atoms in a molecule of nitrogen share three pairs of valence electrons (six valence electrons total) to form a Triple Covalent Bond. Is it possible to predict whether bonds are covalent or not? A good rule of thumb is that bonds between nonmetals (remember that hydrogen is considered a nonmetal) are usually covalent bonds. For example, the carbon dioxide (CO_2) molecules you exhale are bonded together covalently.

Fatal Attraction

Now that you've gotten over being surrounded by covalent bonds, it's time you knew something else. You ate ionic bonds. Lots. Today. You can find some in your saltshaker. In salt (sodium chloride), atoms do not share electrons. Rather, one type of atom strips at least one valence electron from another type of atom, creating ions of opposite charges. These two oppositely charged atoms are held together by an Electrostatic Interaction, an attraction between oppositely charged particles.

An Ionic Bond is an electrostatic interaction that holds together a positively charged ion (cation) and a negatively charged ion (anion). In an ionic bond, one atom loses an electron to another atom, forming a cation and anion, respectively. And, as everyone knows, opposites attract.

In table salt, for example, a valence electron from a sodium atom is transferred to a chlorine atom, forming Na+ and Cl-. Because the ions have opposite charges, they are attracted to each other. The loss of a valence electron and the attraction to the atom that took it happen simultaneously. It is possible for more than one valence electron to be drawn away from another atom, as in barium chloride ($BaCl_2$, a substance used in medicinal preparations). In barium chloride, two chlorine atoms each take one valence electron away from barium, leaving the barium ion Ba^{2+}.

Unlike water or oxygen, which have covalent bonds, substances that have ionic bonds do not exist naturally as discrete molecules. Rather, they form Ionic Solids, three-dimensional networks in which each cation is surrounded by anions and each anion is surrounded by cations.

Look again in the saltshaker. Each sodium cation is surrounded or ionically bonded to six chloride anions, and each chloride anion is ionically bonded to six sodium cations. The formula NaCl for sodium chloride shows that for every sodium atom present in a piece of salt, there is one chlorine atom present. What, don't see any ionic bonds? All right, it looks like this:

Where are ionic bonds found? These types of bonds usually form when metal atoms bond with nonmetal atoms. In salt, the metal sodium bonds with the nonmetal chlorine. Besides salt, some other examples are lithium fluoride (LiF), strontium oxide (SrO) and calcium chloride ($CaCl_2$).

Ules of the Atomic Playground

Atoms, like preschoolers, don't always play fair. This means that sometimes in a covalent bond the electrons are not shared equally between the two atoms. On average, one of the atoms partially "pulls" the bonding electrons toward itself, creating an unequal sharing of those bonding electrons. This is called a Polar Covalent Bond. In order to determine whether a covalent bond is polar or not, it's necessary to understand electronegativity.

Recall that electronegativity is a measure of an atom's ability to draw its bonding electrons to itself. Each element has a numeric value corresponding to its electronegativity. The values used here were devised by Linus Pauling, though there are a few other scales of electronegativity values.

Fluorine was determined to be the most electronegative element and has an electronegativity value of 4.0. Francium, the least electronegative element, has a value of 0.7. It is important to remember a general trend in the periodic table: electronegativity increases from left to right going across a period, and it increases from the bottom to the top of a group. For example, in period 2, nitrogen (group 5A) has an electronegativity value of 3.0, compared to 2.5 for carbon (group 4A). In group 5A, phosphorous (period 3) has an electronegativity value of 2.1, compared to 3.0 for nitrogen (period 2).

In an action that resembles toddlers tugging on a toy, a polar covalent bond occurs when one atom with a higher electronegativity draws the bonding electrons toward itself, pulling those electrons away from the atom with the lower electronegativity value. This creates an unequal sharing of electrons known as Unequal Charge Distribution, or charge separation. The charge separation makes the bond polar because the more electronegative atom becomes partially negatively charged and the atom with the lower electronegativity becomes partially positively charged.

Consider H_2 and HBr. In H_2, the atoms have an equal "pull" on the bonding electrons, making the bond Nonpolar.

In HBr, however, Br has an electronegativity of 2.8, compared to 2.1 for hydrogen. The Br atom pulls the bonding electrons toward itself, creating a partial negative charge on itself and a partial positive charge on the hydrogen atom.

The larger the absolute difference in electronegativity between two atoms, the more polar that bond is. For example, the electronegativity difference of a carbon-oxygen bond is -1.0, the result of 2.5 (the electronegativity value for carbon) minus 3.5 (the value for oxygen). The absolute value for the difference in electronegativity is the value without the minus sign (1.0 for a carbon-oxygen bond). For a carbon-chlorine bond, the difference in electronegativity is 0.5 (2.5-3.0 = -0.5). Thus, a carbon-oxygen bond is more polar (1.0) than a carbon-chlorine bond (0.5).

Sigma Bond

In chemistry, sigma bonds (σ bonds) are the strongest type of covalent chemical bond. Sigma bonding is most clearly defined for diatomic molecules using the language and tools of symmetry groups. In this formal approach, a σ-bond is symmetrical with respect to rotation about the bond axis. By this definition, common forms of sigma bonds are s+s, p_z+p_z, and s+p_z, and $d_z{}^2+d_z{}^2$ (where z is defined as the axis of the bond). Quantum theory also indicates that molecular orbitals (MO) of identical symmetry actually mix. As a practical consequence of this mixing of diatomic molecules, the wavefunctions s+s and p_z+p_z molecular orbitals become blended. The extent of this mixing (or blending) depends on the relative energies of the like-symmetry MO's. For homodiatomics, bonding σ orbitals have no nodal planes between the bonded atoms. The corresponding *anti*bonding, or σ* orbital, is defined by the presence of a nodal plane between these two bonded atoms. Sigma bonds are the strongest type of covalent bonds, and the electrons in these bonds are sometimes referred to as sigma electrons. The symbol σ is the Greek letter for s. When viewed down the bond axis, a σ MO resembles an s atomic orbital.

Sigma Bonds in Polyatomic Compounds

They are obtained by head on overlap of atomic orbitals. The concept of sigma bonding is extended, albeit loosely, to describe bonding interactions involving overlap of a single lobe of one orbital with a single lobe of another. For example, propane is described as consisting of ten sigma bonds, one each for the two C-C bonds and one each for the eight C-H bonds. The σ bonding in such a polyatomic molecule is highly delocalized, which conflicts with the two-orbital, one-bond concept. Despite this complication, the concept of σ bonding is extremely powerful and therefore pervasive.

Sigma Bonds in Multiply Bonded Species

Compounds that feature multiple bonds, such as ethylene and chromium(II) acetate have sigma bonds between the multiply bonded atoms. These sigma bonds are supplemented by π-bonds, e.g. in the case of ethylene, and even δ-bonds, e.g. in the case of chromium(II) acetate.

Sigma Bonds in Organic Molecules

Organic molecules are often made up of one cyclic compound or more, such as benzene, and are often made up of many sigma bonds along with pi bonds. According to the sigma bond rule, the number of sigma bonds in a molecule is equivalent to the number of atoms plus the number of rings minus one.

$$Nb\ \sigma = Nb\ atoms + Nb\ rings - 1$$

This can easily be concluded by realizing that the creation of bonds between atoms that are not connected in a ring requires the same number of atoms minus one (such as in hydrogen gas, H_2, where there is only one sigma bond, or ammonia, NH_3, where there are only 3 sigma bonds), and that rings do not obey this rule (such as benzene rings, which have 6 sigma bonds within the ring for 6 carbon atoms).

Pi Bond

In chemistry, pi bonds (π bonds) are covalent chemical bonds where two lobes of one involved electron orbital overlap two lobes of the other involved electron orbital. Only one of the orbital's nodal planes passes through both of the involved nuclei.

The Greek letter π in their name refers to p orbitals, since the orbital symmetry of the pi bond is the same as that of the p orbital when seen down the bond axis. P orbitals usually engage in this sort of bonding. D orbitals are also assumed to engage in pi bonding but this is not necessarily the case in reality, although the concept of bonding d orbitals still accounts well for hypervalence. Pi bonds are usually weaker than sigma bonds. From the perspective of quantum mechanics, this bond's weakness is explained by significantly less overlap between the component p-orbitals due to their parallel orientation. Pi bonds result from overlap of atomic orbitals that are in contact through two areas of overlap. Pi-bonds are more diffuse bonds than the sigma bonds. Electrons in pi bonds are sometimes referred to as pi electrons. Molecular fragments joined by a pi bond cannot rotate about that bond without breaking the pi bond, because rotation involves destroying the parallel orientation of the constituent p orbitals.

Multiple Bonds

Atoms connected via a double bond or triple bond have, in addition to one sigma bond, one or two pi bonds, respectively.

Although the pi bond by itself is weaker than a sigma bond, pi bonds are often components of multiple bonds, together with sigma bonds. The combination of pi and sigma bond is stronger than either bond by itself. The enhanced strength of a multiple bond versus a single (sigma bond) is indicated in many ways, but most obviously by a contraction in bond lengths. For example in organic chemistry, carbon-carbon bond lengths are in ethane (154 pm), in ethylene (133 pm) and in acetylene (120 pm). More bonds make the total bond shorter and stronger.

Effect of Bond Rotation

Special Cases: Pi bonds do not necessarily connect a pair of atoms that are also sigma-bonded.

In certain metal complexes, pi interactions between a metal atom and alkyne and alkene pi antibonding orbitals form pi-bonds.

In some cases of multiple bonds between two atoms, there is no sigma bond at all, only pi bonds. Examples include diiron hexacarbonyl ($Fe_2(CO)_6$), dicarbon (C_2) and the borane B_2H_2. In these compounds the central bond consists only of pi bonding, and in order to achieve maximum orbital overlap the bond distances are much shorter than expected.

Delta Bond

In chemistry, delta bonds (δ bonds) are chemical bonds of the covalent type, where four lobes of one involved electron orbital overlap four lobes of the other involved electron orbital. Of the orbital's node planes, two (and no more) go through both atoms. The Greek letter δ in their name refers to d orbitals, since the orbital symmetry of the delta bond is the same as that of the usual (4-lobed) type of d orbital when seen down the bond axis.

In chemistry, in sufficiently-large atoms, occupied d-orbitals are low enough in energy to participate in bonding. Delta bonds are usually observed in organometallic species. Some ruthenium and molybdenum compounds contain a quadruple bond, which can only be explained by invoking the delta bond.

It is possible to excite electrons in acetylene from lower-energy nonbonding orbitals to form a delta bond between the two carbon triple bonds. This is because the orbital symmetry of the pi antibonding orbital is the same as that of the delta bond.

Theoretical chemists have conjectured that higher-order bonds (phi bonds and gamma bonds, corresponding to overlap of f and g orbitals) are possible, with even more overlapping lobes of their component atomic orbitals, but no experimental evidence for these has yet been observed.

Double Bond

A double bond in chemistry is a chemical bond between two chemical elements involving four bonding electrons instead of the usual two. The most common double bond, that between two carbon atoms, can be found in alkenes. Many types of double bonds between two different elements exist, for example in a carbonyl group with an carbon atom and an oxygen atom. Other common double bonds are found in azo compounds (N=N), imines (C=N) and sulfoxides (S=O). In skeletal formula the double bond is drawn as two parallel lines between the two connected atoms.

Bonding

The type of bonding can be explained in terms of orbital hybridization. In ethylene each carbon atom has three sp^2 orbitals and one p-orbital. The three sp^2 orbitals lie in a plane with 120 ° angles. The p-orbital is perpendicular to this plane. When the carbon atoms approach each other two of the sp^2 orbitals overlap to form a sigma bond. At the same time the two p-orbitals approach (again in the same plane) and together they form a pi-bond. For maximum overlap the p-orbitals have to remain parallel and therefore rotation around the central bond is not possible. This property gives rise to cis-trans isomerism. Double bonds are shorter than single bonds because p-orbital overlap is maximized.

With 133 pm the C-C bond length is shorter than the C-C length in ethane with 154 pm. The double bond is also stronger, 636 (KJ/mol) versus 368 KJ/mole but not twice as much as the pi-bond is weaker than the sigma bond due to less effective pi-overlap. In an alternative representation the double bond results from two overlapping sp^3 orbitals as in a bent bond.

Variations

In molecules with alternating double bonds and single bonds p-orbital overlap can exist over multiple atoms in a chain giving rise to a conjugated system. Conjugation can be found in systems such as dienes and enones. In cyclic molecules conjugation can lead to aromaticity. In cumulenes two double bonds are adjacent. Double bonds are common for period 2 elements carbon, nitrogen and oxygen and less common with elements of higher periods. Metals too can engage in multiple bonding in a metal ligand multiple bond.

Triple Bond

A triple bond in chemistry is a chemical bond between two chemical elements involving six bonding electrons instead of the usual two in a covalent single bond. The most common triple bond, that between two carbon atoms, can be found in alkynes. Other functional groups containing a triple bond are cyanides and isocyanides. Some diatomic molecules, such as dinitrogen and carbon monoxide are also triple bonded. In skeletal formula the triple bond is drawn as three parallel lines between the two connected atoms..

Bonding

The type of bonding can be explained in terms of orbital hybridization. Each carbon atom has two sp orbitals and two p-orbitals. The two sp orbitals are linear with 180 ° angles and occupy the x-axis (cartesian coordinate system). The p-orbitals are perpendicular on the y-axis and the z-axis. When the carbon atoms approach each other the sp orbitals overlap to form a sp-sp sigma bond. At the same time the two p_z-orbitals approach and together they form a p_z-p_z pi-bond. Likewise, the two p_y-orbitals form a p_y-p_y pi-bond. The result is formation of one sigma bond and two pi bonds.

Quadruple Bond

A quadruple bond is a type of chemical bond between two atoms involving eight electrons. This bond is an extension of the more familiar types double bonds and triple bonds. Stable quadruple bonds are most common among the middle members transition metal elements such rhenium, tungsten, molybdenum and chromium. Typically the ligands that support quadruple bonds are π-donors, not π-acceptors.

Chromium(II) acetate, $Cr_2(\mu\text{-}O_2CMe)_4(H_2O)_2$, was the first chemical compound containing a quadruple bond to be synthesized. It was described in 1844 by E. Peligot, although its distinctive bonding was not recognized for more than a century. The quadruple bond was first characterized in potassium octachlorodirhenate(III) or $K_2[Re_2Cl_8]\cdot 2H_2O$ by F. A. Cotton in 1964. The rhenium-rhenium bond length in this compound is only 224 pm. In the terminology of molecular orbital theory, the bonding is described as $\sigma^2\pi^4\delta^2$ with one sigma bond, two π bonds and one delta bond.

Many other compounds with quadruple bonds have been described, often by Cotton and his coworkers. Isoelectronic with the dirhenium compound is the salt $K_4[Mo_2Cl_8]$ (potassium octachlorodimolybdate). An example of a ditungsten compound with a quadruple bond is di-tungsten tetra(hpp).

Quintuple Bond

A quintuple bond in chemistry is an unusual type of chemical bond first reported in 2005 for a dichromium compound. Single bonds, double bonds, and triple bonds are commonplace in chemistry. Quadruple bonds are rarer but occur especially for Cr, Mo, W, and Re, e.g. $[Mo_2Cl_8]^{4-}$ and $[Re_2Cl_8]^{2-}$. In a quintuple bond, ten electrons participate in bonding between the two metal centers, allocated as $\sigma^2\pi^4\delta^4$.

In some cases of high-order bonds between metal atoms, the metal-metal bonding is facilitated by ligands that link the two metal centers and reduce the interatomic distance. By contrast, the chromium dimer with quintuple bonding is stabilized by bulky terphenyl (2,6-[(2,6-diisopropyl)phenyl]phenyl) ligands.

The species is stable up to 200 °C. The chromium-chromium quintuple bond has been analyzed with multireference ab-initio and DFT methods, which were also used to elucidate the role of the terphenyl ligand, in which the flanking aryls were shown to interact very weakly with the chromium atoms, causing only a small weakening of the quintuple bond. A 2007 theoretical study identified two global minima for quintuple bonded RMMR compounds: a trans-bent molecular geometry and surprisingly another trans-bent geometry with the R substituent in a bridging position.

In 2005, a quintuple bond was postulated to exist in the hypothetical uranium molecule U_2 based on computational chemistry. Diuranium compounds are rare, but do exist; for example, the U_2Cl2 8 anion.

In 2007 the shortest ever metal to metal bond (180.28 pm) was reported to exist also in a compound containing a quintuple chromium-chromium bond with diazadiene bridging ligands. Other metal-metal quintuple bond containing complexes that have been reported include quintuply bonded dichromium with [6-(2,4,6-triisopropylphenyl) pyridine-2-yl] (2,4,6-trimethylphenyl) amine bridging ligands and a dichromium complex with amidinate bridging ligands.

Synthesis of quintuple bonds is usually achieved through reduction of a dimetal species using potassium graphite. This adds valence elections to the metal centers, giving them the needed number of electrons to participate in quintuple bonding.

Bonding

As stated above metal-metal quintuple bonds have a $\sigma^2\pi^4\delta^4$ configuration. This means the of the five bonds present between the metal centers one is a sigma bond, two are pi bonds, and two are delta bonds. The σ-bond is the result of mixing between the $d_z{}^2$ orbital on each metal centre. The first π-bond comes from mixing of the d_{yz} orbitals from each metal while the other π-bond comes from the d_{xz} orbitals on each metal mixing. Finally the δ-bonds come from mixing of the d_{xy} orbitals as well as mixing between the $d_x{}^2{}_{-y}{}^2$ orbitals from each metal. A graphical representation of this mixing can be seen below.

Ligand Role in Metal-metal Quintuple Bond Length

Quintuple bond lengths vary from complex to complex and are heavily dependent on the ligands bound to the metal centers. Nearly all complexes containing a metal-metal quintuple bond have bidentate bridging ligands, and even those that don't like the terphenyl complex mentioned earlier, have some bridging characteristic to it through metal-ipso-carbon interactions.

The bidentate ligand can act as a sort of tweezer in that in order for chelation to occur the metal atoms must move closer together, thereby shortening the quintuple bond length. The two ways in which to obtain shorter metal-metal distances is to either reduce the distance between the chelating atoms in the ligand by changing the structure, or by using steric effects to force a conformational change in the ligand that bends the molecule in a way that forces the chelating atoms closer together. An example of the latter is shown below:

Future Prospects

Research in quintuple bonds is still in its infancy and as a result is more of an interesting phenomenon that has no practical applications as of yet. These quintuple bond containing organometallic compounds don't seem to lend themselves for use as catalysts due to their lack of coordination sites.

Current research is centered mostly on making shorter quintuple bonds, though new reactions have been reported that have lead to rarely seen metal-ligand interactions. This suggests metal-metal quintuply bonded complexes could be used as precursors to make previously unattainable compounds.

Sextuple Bond

A sextuple bond is a type of covalent bond involving 12 bonding electrons and in which the bond order is 6. The only known molecules with true sextuple bonds are the diatomic dimolybdenum (Mo_2) and ditungsten (W_2), which exist in the gaseous phase. There is strong evidence to believe that no two elements in the periodic table can form a bond with greater order than 6.

Dimolybdenum and Ditungsten

Dimolybdenum (Mo_2) can be observed in the gas phase at low temperatures (7K) by a laser evaporation technique using molybdenum sheet with, for instance, near infrared spectroscopy or UV spectroscopy. Like dichromium, a singlet state is expected from dimolybdenum. Higher bond order is reflected in shorter bond length of 194 pm.

Other Molecules

Although diatomic Cr_2 and U_2 have formal structures with twelve-electron bonds, their effective bond orders derived from quantum chemistry calculations are less than 5 (which would be quintuple bonds).

Factors Influencing Corrosion

In any discussion of the mechanism of a chemical reaction it is advisable to separate the factors which determine the tendency or driving force of the reaction to proceed from those which influence the rate of the reaction made possible by the existence of this tendency. This tendency is an expression of the fact that the system is not in a state of equilibrium (or inherent stability); it is measured by the difference in energy between the initial and final state of the system for any particular case. In most cases the observed rate is determined not by the absolute magnitude of this tendency but by other factors, which depend primarily upon the environment.

In considering the group of three typical reactions involved in corrosion, we shall denote as primary factors those which determine the tendency of the metal to corrode and thus influence its initial rate of solution and as secondary factors those which influence the rate of the subsequent reactions. This term in no wise implies that these secondary factors are of lesser importance; in fact, by influencing the nature and distribution of the final corrosion products, they usually determine the ultimate rate of corrosion, and the useful life of the metal, in each environment. In the general case, some one or two of the many factors involved exert outstanding influence upon the ultimate rate of corrosion; these we term controlling or dominant factors. In general, the primary factors have to do with the metal (or alloy) itself; the secondary factors more with the specific environment. It is convenient to divide them in this way, although no sharp distinction can be made.

Factors Associated with Metals

- Effective electrode potential of a metal in a solution
- Over voltage of hydrogen on the metal
- Chemical and physical homogeneity of the metal surface
- Inherent ability to form an insoluble protective film

Factors Which Vary Mainly with the Environment

- Hydrogen-ion concentration (pH) in the solution
- Influence of oxygen in solution adjacent to the metal
- Specific nature and concentration of other ions in solution
- Rate of flow of the solution in contact with the metal
- Ability of environment to form a protective deposit on the metal
- Temperature
- Cyclic stress (corrosion fatigue)
- Contact between dissimilar metals or other materials as affecting localized corrosion.

Electrolytic corrosion consists of two partial processes: an anodic (oxidation) and cathodic (reduction) reaction. In the absence of any external voltages, the rates of the anodic and cathodic reactions are equal, and there is no external flow of current.

Oxidation Reaction

This reaction represents the oxidation of a metal (M) from the elemental (zero valence) state to an oxidation state of M^{n+} with the generation of n moles of electrons (e^-). The anodic reaction may occur uniformly over a metal surface or may be localized to a specific area. If the dissolved metal ion can react with the solution to form an insoluble compound, then a buildup of corrosion products may accumulate at the anodic site.

In the absence of any applied voltage, the electrons generated by the anodic reaction are consumed by the cathodic reaction. For most practical situations, the cathodic reaction is either the hydrogen-evolution reaction or the oxygen-reduction reaction. The hydrogen-evolution reaction can be summarized as reaction.

$$2H^+ + 2e^- \rightarrow H_2$$

In this case, protons (H^+) combine with electrons to form molecules of hydrogen (H_2). This reaction is often the dominant cathodic reaction in systems at low pH. The hydrogen-evolution reaction can itself cause corrosion-related problems, since atomic hydrogen (H) may enter the metal, causing embrittlement, a phenomenon that results in an attenuation of the mechanical properties and can cause catastrophic failure.

The second important cathodic reaction is the oxygen-reduction, given by reactions.

$$O_2 + 4H^+ + 4e^- \rightarrow 2H_2O$$

$$O_2 + 2H_2O + 4e^- \rightarrow 4OH^-$$

These represent the overall reactions in acidic and alkaline solutions, respectively. This cathodic reaction is usually dominant in solutions of neutral and alkaline pH. In order for this reaction to proceed, a supply of dissolved oxygen is necessary; hence the rate of this reaction is usually limited by the transport of oxygen to the metal surface.

Reactions above represent the overall reactions which, in practice, may occur by a sequence of reaction steps. In addition, the reaction sequence may be dependent upon the metal surface, resulting in significantly different rates of the overall reaction. The cathodic reactions are important to corrosion processes since many methods of corrosion control depend on altering the cathodic process. Although the cathodic reactions may be related to corrosion processes which are usually unwanted, they are essential for many applications such as energy storage and generation.

Corrosion rates are usually expressed in terms of loss of thickness per unit time. General corrosion rates may vary from on the order of centimeters per year to micrometers per year. Relatively large corrosion rates may be tolerated for some large structures, whereas for other structures small amounts of corrosion may result in catastrophic failure. For example, with the advent of technology for making extremely small devices, future generations of integrated circuits will contain components that are on the order of nanometers ($10''^{9}$ m) in size, and even small amounts of corrosion could cause a device failure. In some situations, corrosion may occur only at localized regions on a metal surface. This type of corrosion is characterized by regions of locally severe corrosion, although the general loss of thickness may be relatively small.

Pitting corrosion is usually associated with passive metals, although this is not always the case. Pit initiation is usually related to the local breakdown of a passive film and can often be related to the presence of halide ions in solution.

Crevice corrosion occurs in restricted or occluded regions, such as at a bolted joint, and is often associated with solutions that contain halide ions. Crevice corrosion is initiated by a depletion of the dissolved oxygen in the restricted region. As the supply of oxygen within the crevice is depleted, because of cathodic oxygen reduction, the metal surface within the crevice becomes activated, and the anodic current is balanced by cathodic oxygen reduction from the region adjacent to the crevice. The ensuing reactions within the crevice are the same as those described for pitting corrosion: halide ions migrate to the crevice, where they are then hydrolyzed to form metal hydroxides and hydrochloric acid.

Corrosion can also be accelerated in situations where two dissimilar metals are in contact in the same solution. This form of corrosion is known as galvanic corrosion. The metal with the more negative potential becomes the anode, while the metal with the more positive potential sustains the cathodic reaction. In many cases the table of equilibrium potential can be used to predict which metal of galvanic couple will corrode. For example, aluminum-graphite composites generally exhibit poor corrosion resistance since graphite has a positive potential and aluminum exhibits a highly negative potential. As a result, in corrosive environments the aluminum will tend to corrode while the graphite remains unaffected.

Stress corrosion cracking and hydrogen embrittlement are corrosion-related phenomena associated with the presence of a tensile stress. Stress corrosion cracking results from a combination of stress and specific environmental conditions so that localized corrosion initiates cracks that propagate in the presence of stress. Mild steels are susceptible to stress corrosion cracking in environments containing hydroxyl ions (O≈; often called caustic cracking) or nitrate ions (NO_3"). Austenitic stainless steels are susceptible in the presence of chloride ions (Cl") and hydroxyl ions (O≈). Other alloys that are susceptible under specific conditions include certain brasses, aluminum and titanium alloys. Hydrogen embrittl-ement is caused by the entry of hydrogen atoms into a metal or alloy, resulting in a loss of ductility or cracking if the stress level is sufficiently high. The source of the hydrogen is usually from corrosion (that is, cathodic hydrogen evolution) or from cathodic polarization. In these cases the presence of certain substances in the metal or electrolyte can enhance the amount of hydrogen entry into the alloy by poisoning the formation of molecular hydrogen. Metals and alloys that are susceptible to hydrogen embrittlement include certain carbon steels, high-strength steels, nickel-based alloys, titanium alloys, and some aluminum alloys. A reduction in the rate of corrosion is usually achieved through consideration of the materials or the environment. Materials selection is usually determined by economic constraints. The corrosion resistance of a specific metal or alloy may be limited to a certain range of pH, potential, or anion concentration. As a result, replacement metal or alloy systems are usually selected on the basis of cost for an estimated service lifetime.

Different Types Corrosion

Wet Corrosion

Wet corrosion of metals occurs through electron transfer, involving two processes, oxidation and reduction. In oxidation, the metal atoms lose electrons. The surrounding environment then gains the electrons in reduction. The metal, where electrons are lost, is called the anode. The other metal, liquid or gas which gains the electrons is called the cathode.

Corrosion Resistance of Aluminum

The corrosion resistance of aluminium varies widely depending on alloy, environment, design and protective measures taken. However, it is possible to give some general guide-lines. A clean aluminium surface is reactive and will react spontaneously with water or air and form aluminium oxide. This oxide is very stable and has in addition a very good adhesion to the metal surface and thus protects aluminium from corrosion or further oxidation. This means that aluminium has good corrosion resistance in environments where the oxide layer is stable.

The oxide layer will deteriorate in environments with high or low pH or where aggressive ions are present. Below a pH of 4, acidic conditions, and above a pH of 8,5, alkaline conditions, there will normally be an increase in the corrosion rate of aluminium, but this also depends on which ions that are present in the environment.

Aggressive ions will break down the oxide layer locally and start local corrosion attacks. Among the aggressive ions, chloride (Cl-), is the one with the most practical importance, because it is present in large amounts in both sea-water, road salts and some soils and in lower concentrations in other natural sources.

A so called general corrosion attack proceeds at about the same rate on the entire metal surface. Because of the stability of the oxide layer, general corrosion will rarely be a problem on aluminium, except in very alkaline or acidic environments. Aluminium may however experience local attacks in connection with formation of small anodic areas as a result of a local breakdown of the oxide layer. Some of the most typical local attacks on aluminium are pitting corrosion, crevice corrosion, intergranular corrosion and galvanic corrosion.

Types of Corrosion

When there is a relative movement between the corrosive liquid and the metal surface, the metal may be exposed to mechanical wear which removes the protective oxide layer and enhances the corrosion. This is called flow influenced corrosion and can be further divided into;

Erosion- and Turbulence-corrosion

Pure erosion is caused by a shear force from a flowing liquid which is higher than the adhesion of the oxide layer on the aluminium surface. The shear force from the liquid may be increased by turbulence or dissolved or suspended solids causing abrasion. The shear force from the liquid is not necessarily caused by liquid flow, but can also be induced by a moving surface for example a propeller.

If the removal of the oxide layer occurs in a corrosive liquid, the corrosion will be enhanced because a bare aluminium surface will be exposed to the liquid for a while until the oxide layer is healed. This is called erosion-corrosion. The rapid self healing of the oxide layer on aluminium is a great advantage with regards to erosion-corrosion, but because aluminium is a fairly soft material, erosion-corrosion can be a serious problem. The shear force from the liquid on the metal surface is the main design criteria with regards to erosion corrosion. In practice, this is often transferred to flow rate limits. For the same flow rate, the shear force in a small diameter tube will be higher than in a large diameter tube, but the differences will be small for normal flow rates. If the liquid contains no solids and in smooth pipes, with undisturbed flow, the tolerable flow rate will be fairly high, but normally some turbulence must be expected and the flow rate limitations must be decided, based on these critical areas.

The data on flow rate limitations on aluminium are rather limited. An upper limit on 2ρ3 m/s in flowing sea water is often suggested. In addition to limiting the flow rates, it is very important to avoid introducing turbulence sources like for example sharp bends and irregularities around joints.

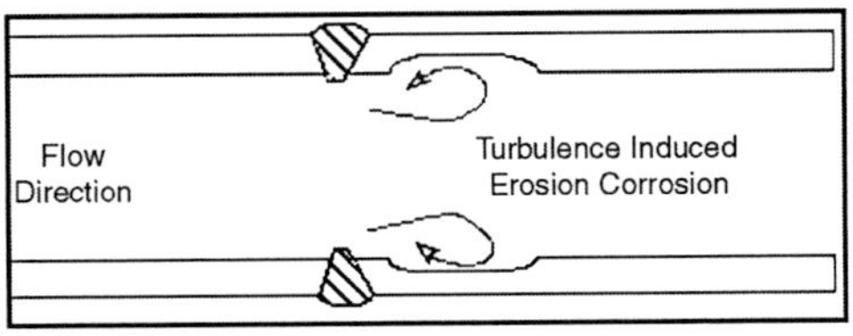

Impingement Corrosion

Impingement corrosion can occur for example in bends if the liquid contains gas bubbles or solids, or if a gas flow contains droplets. The droplets or bubbles will hammer against the metal surface and can cause rapid failure if the conditions are unfavourable. Impingement in a bend, arrows show the direction of the main flow.

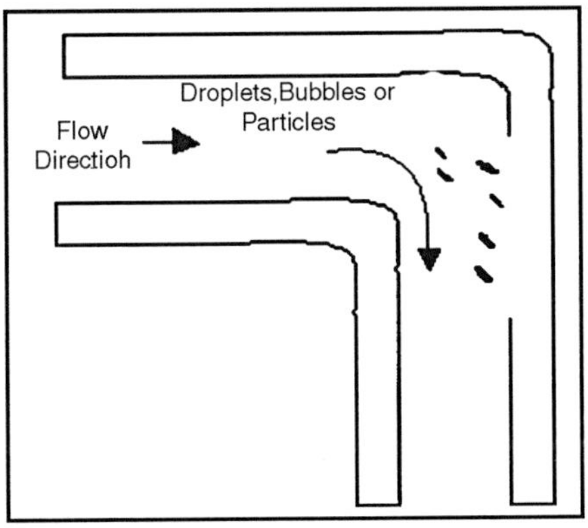

If there is a risk for bubbles or droplets in the flow, it is very important that the flow rate is kept low and bends are made smooth and not as sharp 90*f* angles. It is also desirable to separate out solids, water or gas early in a flow system in order to avoid two phase flow.

Cavitations Corrosion

Cavitations corrosion occurs in areas with high flow rates and rapid pressure changes. This causes gas bubbles to collapse.

This gives a high force on the metal surface which can remove oxide and give an active aluminium surface. Good design to avoid large pressure changes or hard surface layers is ways to avoid cavitations corrosion.

Pitting Corrosion

Pitting corrosion is the most common corrosion form seen on aluminium. It starts from a weak point in the oxide, and after initiation, the reactions within the pit will reduce the pH and increase the chloride concentration and in this way sustain the reaction. Pitting can penetrate several mm during a short period if the conditions are extremely unfavourable.

Normally the development of pit depths follows a time function of the form:

$$D = K \cdot tc$$

Where D is the depth of the deepest pit, K is an alloy-, surface area- and environment-dependent constant, t is time and c is an environment-dependent parameter which often is close to 1/3. For good alloys immersed in sea water, with t given in years and D in mm, K can typically be close to 0,75. The development of the maximum pit depth will typically follow the form shown below.Typical development of pitting corrosion rate, the figures on the axes is only examples. With a rapid increase after initiation followed by a slower growth, the pits can be of different shapes, wide or narrow. The narrow pits can often be dangerous because the attack can be deep and difficult to detect.

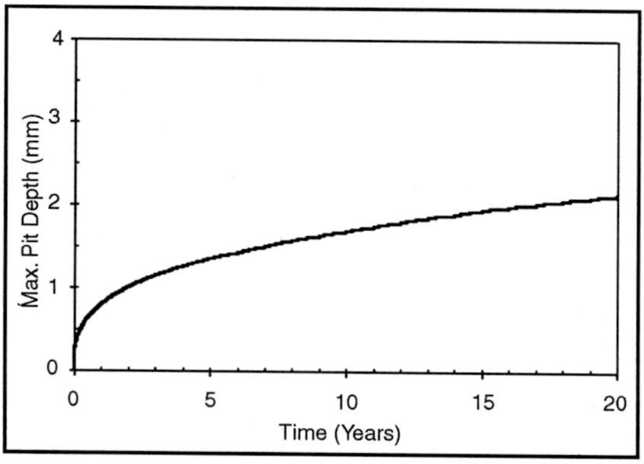

Galvanic Corrosion

When two metallic materials are in contact in the presence of a corrosive electrolyte, corrosion will increase on the least noble material (the anode) and decrease on the noblest (the cathode). The increase in corrosion is called galvanic corrosion.

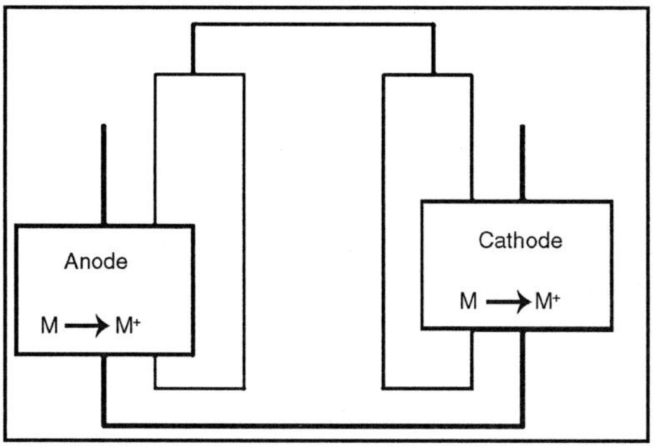

Galvanic corrosion occurs when two different metals are connected in an electrolyte and one metal corrodes in protection of the other. One metal is corroding in the same manner as we saw in wet corrosion ($M \rightarrow M^{n+} + ne^-$) while the other metal is reacting ($M^{n+} + ne^- \rightarrow M$).

From these two reactions we can see that the anodic electrode corrodes and the cathodic electrode is protected. The corroded metal is called the sacrificial anode. Since aluminium is less noble than most commonly used construction materials, with the exception of zinc, cadmium and magnesium, this can be a serious corrosion form with aluminium. The protective effect from the oxide layer can be seriously deteriorated by the coupling to a nobler material, this is especially dangerous in atmospheres or waters with high concentrations of chlorides or other aggressive species.

There is relatively small difference in the galvanic corrosion rate for different aluminium alloys, because the corrosion rate is largely decided by the cathodic properties of the noblest material. A galvanic connection will very often amplify exfoliation, intergranular and stress corrosion in alloys susceptible to these corrosion forms.

Crevice Corrosion

Crevice corrosion occurs in narrow metal to metal or non-metal to metal gaps where the convection of water is hampered and a specific crevice chemistry different from the environment is allowed to develop. Aggressive ions like chlorides must be present in the electrolyte. The oxygen in the bottom of the crevice is consumed and an anodic area is developed adjacent to the oxygen depleted zone. Crevice corrosion develops quite similar to pitting corrosion after the initiation stage, with a gradual decrease of the pH and an increase of the chloride concentration within the crevice. Crevice corrosion is normally not a serious problem on aluminium in the absence of aggressive ions, because of the very stable aluminiumoxide. However in a crevice there will be a possibility for accumulation of moisture because of capillary forces and deposits with corrosive or hygroscopic species. In this way there will be a constant corrosive environment in the crevice which eventually can break down the oxide layer.

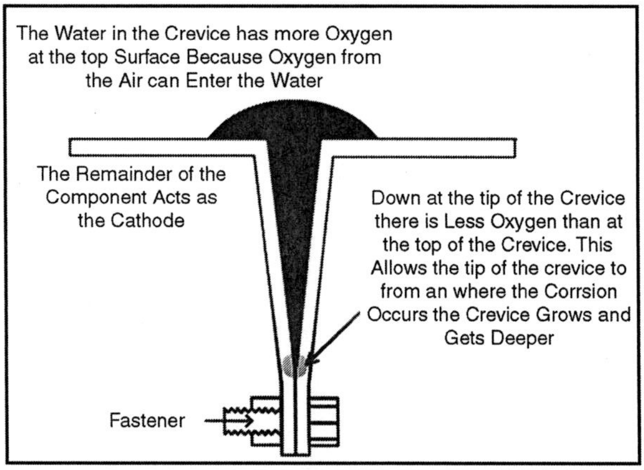

Crevice corrosion can occur during storage of aluminium such as water-staining which creates dark stains as a result of a surface etching caused by water trapped between the adjacent surfaces. It is most commonly seen on sheet products and is mainly an aesthetic problem as the mechanical integrity of the water stained aluminium is rarely impaired. Another possible problem is accumulation of voluminous corrosion products in narrow crevices which can finally deform the construction in extreme cases.

Intergranular Corrosion

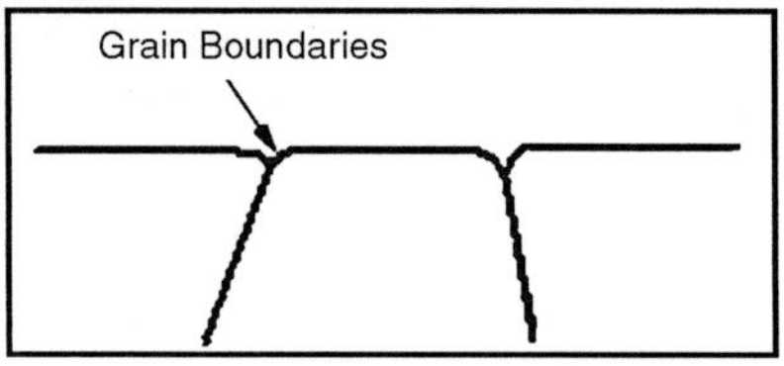

Intergranular corrosion proceeds along the grain boundaries. This form of corrosion can be very dangerous because the attack can proceed undetected through the material and may deteriorate the mechanical properties of the metal and cause fracture without any visible exterior signs of corrosion. Susceptibility to intergranular corrosion is mainly dependent upon alloy composition and heat-treatment. The alloys in the 6000-group are normally very resistant to intergranular corrosion as long as the Mg/Si ratio does not increase too much above the stoichiometric ratio of 1,73, the copper content should be kept at a low level and correct heat-treatment procedures should be adhered to.

In addition the heat affected zone of a weld that is not properly executed, may be a starting point for intercrystalline corrosion. Internal weld zones in the profile may also be a starting point for preferential corrosion attacks if there is associated segregation of alloying elements. This is quite similar to intergranular corrosion and these welds should be placed in non-critical areas of the profile.

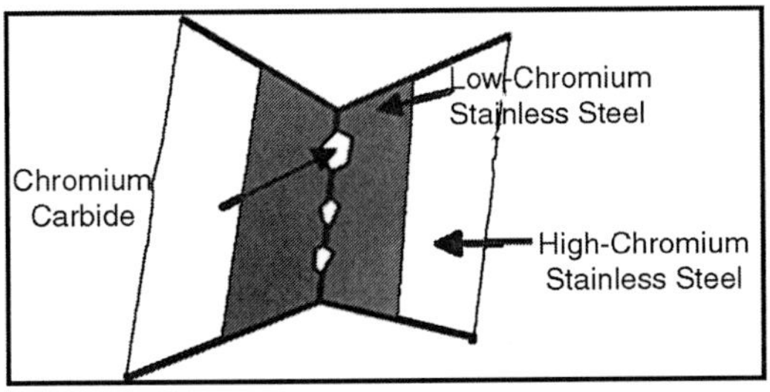

In stainless steels the problem occurs after the metal is heated to between 425°C and 870°C. During the heating, the chromium in the stainless steel reacts with carbon in the steel and forms particles of chromium carbide at the grain boundaries. The regions near the grain boundaries become depleted in chromium.

Stress Corrosion

Stress corrosion is another form of corrosion that is important to many fields including civil structures. Stress-corrosion occurs when a material exists in a relatively inert environment but corrodes due to an applied stress. The stress may be externally applied or residual.

This form of corrosion is particularly dangerous because it may not occur under a particular set of conditions until there is an applied stress. The corrosion is not clearly visible prior to fracture and can result in catastrophic failure.

Many alloys can experience stress corrosion, and the applied stress may also be due to a residual stress in the material. An example of a residual stress could be a stress remaining in a material after forming, or a stress due to welding. Stress corrosion cracking will usually cause the material to fail in a brittle manner, which can have grave consequences as there is usually little or no warning before the failure occurs. Stress corrosion is a form of galvanic corrosion, where stressed areas of the material are anodic to the unstressed areas of the material. Practically the best way to control stress corrosion cracking is to limit or reduce the stresses a material is under while it is in a corrosive atmosphere.

Dry Corrosion

At room temperature, most metals carry a thin oxide layer as a result of the reaction of metals with oxygen in the atmosphere. Increase of temperature may cause formation of a heavier layer, or the layer may detach. Zinc and zinc coatings carry a fairly protective zinc hydroxide or carbonate layer (zinc patina) which increases in thickness very slowly. Aluminium carries a thin, highly protective oxide layer. Some corrosion takes place even under completely dry conditions.

The basic reaction involved in dry corrosion is:

$$M \Rightarrow M^{n+} + n \times e^{-1}$$

Where M is a metal element.

The metal loses electrons to form an ion and some free electrons. The ionic metal can then react with oxygen to form a metal oxide. In dry corrosion, the oxygen comes from the air, in wet corrosion, the oxygen is supplied by aerated water

Recently, we invented novel nanocrystalline metal-base nanocomposites distributing nanoparticles of Cr or/and Al through co-electrodeposition of metal matrix with nanoparticles These nanocomposites containing appropriate contents of Cr or/and Al can quickly form protective films, when they are suffered the attack of corrosive species at ambient or high temperatures. Hence, they exhibit excellent corrosion resistance, as compared to conventionally coarse-grained (CG) alloys

These kinds of nanocomposites are proposed as surface coatings for metals against liquid corrosion hot corrosion and high temperature oxidation Moreover, the nanocomposites contain numerous grain boundaries (GBs); this causes highly enhanced diffusion kinetics of atoms in them with respect to in the CG alloy counterparts. For this reason, the nanocomposites may be applied to metals as precursor films for chemical heat treatment such as nitridation to further improve their surface properties for various proposes. This improvement of surface properties would be more significantly, if the chemical treatment on the nanocomposites is conducted at temperatures greatly lower than normal temperatures adopted by similar treatment on conventional CG alloys. In this paper, the fabrication, wet and dry corrosion behaviours, and plasma nitridation performance of the types of electrodeposited nanocomposites (ENCs) are we briefly reviewed taking ENC Ni-Cr as an example.

Corrosion in Dry Systems

Mother Nature has the task of reducing compounds down to their primary constitutes, and this is accomplished very effectively. Corrosion of fire sprinkler systems is inevitable. One of the challenges faced by the fire sprinkler industry is to minimize this process. This article seeks to explain that there is not a single total solution to corrosion control in fire sprinkler systems. Instead, our industry must battle corrosion on several different fronts.

Atmospheric Corrosion

Components exposed to the elements will inevitably experience damage due to atmospheric corrosion. The severity of the corrosion and the rate at which corrosion will take place are dependant primarily upon the properties of the surface formed electrolytes, which in turn are dependant upon factors such as the humidity and pollution levels in the atmosphere.

Assessing the corrosivity of a particular atmosphere can be relatively simple. Two methods have been developed to deal with this problem: The first method involves the exposure of metallic coupons to the environment and classifying the resultant corrosion, *e.g.* the coupon based method (CLIMAT) based on the wire on bolt ASTM,

$$\text{Anode Reaction}: 2M \rightarrow 2M^{2+} + 4e^-$$

$$\text{Cathode Reaction}: O_2 + 2H_2O + 4e^- \rightarrow 4OH^-$$

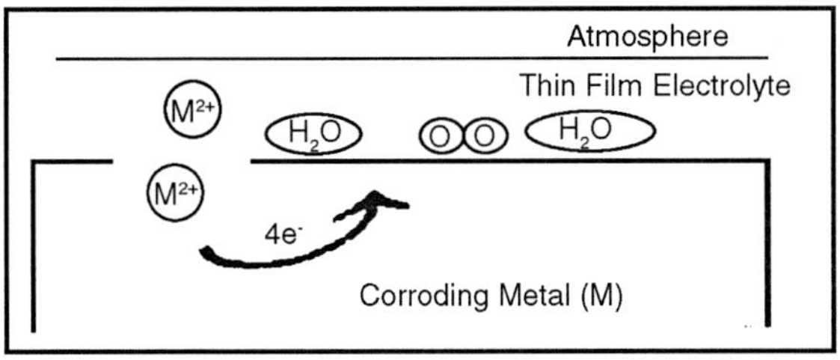

Atmospheric corrosion occurs on a steel surface in a thin wet film created by the humidity in the air in combination with impurities.

Corrosive conditions for stainless steel can be the chloride depositions from a marine atmosphere on the steel surface in a humid atmosphere these, conditions may alter the choice of steel grade to a more corrosion resistant alloy. The type of corrosion is most commonly local in the presence of chlorides, *i.e.* pitting or crevice corrosion. Atmospheric environments are most commonly divided into four categories: rural, urban, industrial and marine. The environments vary depending on the severity from a corrosive point of view. The importance of keeping the surface clean by regular washing to avoid staining and dust cannot be stressed enough.

Mechanism

Atmospheric corrosion is an electrochemical process, requiring the presence of an electrolyte. Thin film "invisible" electrolytes tend to form on metallic surfaces under atmospheric corrosion conditions, when a certain critical humidity level is reached. For iron, this level is around 60%, in unpolluted atmospheres. The critical humidity level is not a constant - it depends on the corroding material, the hygroscopic nature of corrosion products and surface deposits and the presence of atmospheric pollutants.

Monitoring

Atmospheric corrosion accounts for the highest overall cost and metal loss of all the fundamental corrosive environments. A defining feature of atmospheric corrosion is the thin aqueous layer between the surface of the corroding material and the atmosphere.

Three phases (solid [corroding substrate], liquid [thin aqueous layer] and gaseous[atmosphere]) and the interfaces between these phases are therefore important and can be used in corrosion monitoring principles..Corrosion monitoring in outdoor and indoor atmospheres poses specific challenges related to characterizing corrosion damage (generally taking place at a low rate) in a short (practical) time frame.

Direct Measurement

(Examples include exposing actual components or coated coupons to corrosive atmospheres and evaluating these for corrosion damage periodically)

Classification

By categories and correlating such classifications to actual in-service performance and corrosion rates. The main drawbacks of the direct measurement approach are the lengthy exposure time period usually required and that only a “snapshot” of cumulative damage is obtained when detailed analysis of corrosion damage is performed periodically.

The approach of atmospheric corrosivity classification is generally one of a simple, low-cost measurement in a short time frame. Ultimately, such simplistic measurements require correlation to actual long term service performance and therefore need to be linked to the other measurement methodologies. Atmospheric corrosion monitoring is generally employed to quantify this type of damage and corrosion risk, rather than merely resorting to broadly descriptive atmospheric classifications such as “industrial”, “rural”, etc.

Soil Corrosion

Soil corrosion is a complex phenomenon, with a multitude of variables involved. Chemical reactions involving almost each of the existing elements are known to take place in soils, many of which are not yet fully understood. The relative importance of variables changes for different materials, making a universal guide to corrosion impossible. Variations in soil properties and characteristics across three dimensions can have a major impact on corrosion of buried structures.

In the presence of thin film electrolytes, atmospheric corrosion proceeds by balancing anodic and cathodic reactions. The anodic oxidation reaction involves the dissolution of the metal in the electrolyte, while the cathodic reaction is often assumed to be the oxygen reduction reaction. Oxygen from the atmosphere is readily supplied to the electrolyte, under thin film corrosion conditions.

Variations in soil properties and characteristics across three dimensions can have a major impact on corrosion of buried structures.

The response of carbon steel to soil corrosion depends primarily on the nature of the soil and certain other environmental factors, such as the availability to moisture and oxygen. These factors can lead to extreme variations in the rate of the attack. For example, under the worst condition a buried vessel may perforate in less than one year, although archaeological digs in arid desert regions have uncovered iron tools that are hundreds of years old. (Reference)

Some general rules can be formulated. Soils with high moisture content, high electrical conductivity, high acidity, and high dissolved salts will be most corrosive. The effect of aeration on soils is somewhat different from the effect of aeration in water because poorly aerated conditions in water can lead to accelerated attack by sulfate-reducing anaerobic bacteria.

The effect of low levels of alloying additions on the soil corrosion of carbon steels is modest. Some data seems to show a small benefit of 1%Cu and 2.5% Ni on plain carbon steel. The weight loss and maximum pit depth in soil corrosion can be represented by an equation of the form:

$$Z = a \cdot t^m$$

where:

- *Z*: Either the weight of loss of maximum pit depth
- *T*: Time of exposure
- *A and m*: Constants that depend on the specific soil corrosion situation.

Soil Classification

Soil texture refers to the size distribution of mineral particles in a soil. Sand (rated from coarse to very fine), silt and clay refer to textures of decreasing particle coarseness. Soils with a high proportion of sand have very limited storage capacity for water, whereas clays are excellent in retaining water. One soil identification system has defined eleven soil types on the basis of their respective proportions of clay, silt and sand.

The Ten types are:

1. Sand
2. Loamy sand
3. Sandy loam
4. Sandy clay loam
5. Clay loam
6. Loam
7. Silty loam
8. Silt
9. Silty clay loam
10. Silt clay and clay

Another identification scheme has utilized chemical composition, organic content and history of formation to define types such as gravel, humus, marsh and peat. A newer soil classification system has evolved in the country that can be utilized to classify soils globally, at any location. In this "universal" classification system soils are considered as individual three- dimensional entities that can be grouped according to similar physical, chemical and mineralogical properties.

Measure Against Corrosion

Protection of Steel

Steel reinforcements are the backbone of any concrete structure and provide tensile strength to complement the compressive strength of concrete. Together they provide strength to the structure. Concrete under ideal conditions is maintenance free and has almost infinite life. However, the presence of steel makes it vulnerable to the attack of corrosion and in the process reduces the useful, maintenance free life of the RCC structures. The corrosion of steel used in buildings, bridges and other concrete structures (especially in coastal and marine environment) in the form of reinforcement bars leads to the deterioration of concrete. This in turn reduces the load bearing capacity of concrete thereby rendering it vulnerable to failure. This effectively creates the need for undertaking expensive maintenance and restoration work, the cost of which at times runs into many times the original cost of construction.

This is more severe and conspicuous in the coastal regions. Hence there is an urgent need to protect the steel against corrosion, especially in the future projects coming up across the country accounting for crores of rupees in investments and ensure that the useful life of these projects are enhanced substantially to serve the nation with many more years of service life.

Corrosion induced damage to reinforced concrete often necessitates early repair and occasionally complete replacement of the structure or element well before its design life is reached. Worldwide, the costs associated with such remedial work are massive and are expected to increase in the future at an alarming rate. Estimates of these costs vary widely. For example, as early as 1977 in the USA, the cost of major repair and replacement of damaged bridge decks, which was rapidly becoming a systemic problem, was about $23 billion.

Broomfield reported that in the late 1980s the UK Department of Transport estimated repair costs of some £616 million for motorway bridges in England and Wales. Over the same period in the USA, annual costs associated with bridge deck repairs due to deicing salts alone was in the range 50-200 million, plus an additional 100 million for substructures and 50-100 million on multi-level parking garages. India has a large coastal belt starting from Gujarat, Goa and Maharashtra in the west to Tamil Nadu, Karnataka and Kerala in the south, and Andhra Pradesh, West Bengal and Orissa to the east. It has one of the largest coastlines and highly aggressive environments, leading to accelerated corrosion of the steel being used in these regions. The Indian Lead Zinc Development Association and CECRI figures put the total annual loss due to corrosion at an amount equal to 2 per cent of the GDP of nation.

Clearly, such costs are an enormous drain on finances and resources which could otherwise be allocated to new construction. What has also become evident is that while the repair of reinforced concrete may make good the surface deterioration of the structure, it usually does not address the root causes of the problem.

Corrosion Rates

The corrosion map of India developed by the Corrosion Advisory Bureau, NML (CSIR), indicates the corrosion rate across the country. The different regions are classified into severe corrosion, high corrosion, moderate corrosion and least corrosion and minimal corrosion belts, depending on the local environment. It is seen that a major part of the coastal regions including Chennai, Mumbai, Kolkata and Visakhapatnam fall in the severe corrosion area. These regions especially require precautions to be taken to prevent corrosion of steel thereby necessitating the need for steel protection. The remaining areas would not require this level of protection.

Thus there is a pressing need for government organisations, project bodies, consultants and contractors in the corrosion-prone areas to come tighter and take concerted efforts to protect steel reinforcements from deterioration and thus protect the construction from premature failure. Hence the need to evaluate appropriate and user-friendly technologies for protecting steel and decide on the form of protection to be given to rebars.

Protection Measures

Corrosion of steel reinforcement in concrete is an electrochemical process which requires access of an electrolyte and oxygen to steel. Rust-stained surfaces and cracking and spelling of concrete in recently completed structures demonstrate the imperative need to protect steel reinforcement. Protective measures against corrosion rely on minimizing or preventing the corrosion (electrochemical) process.

Four types of protective measures can be identified:

1. Impeding access of deleterious materials (water, oxygen, salts, carbon dioxide etc.) to the steel surface.
2. Slowing the electrochemical process through use of inhibitors.
3. Modifying the electrode through cathodic protection.
4. Providing coatings to the steel reinforcement.

Out of the above mentioned alternatives, coating of rebars has been proved to be the best one and galvanizing has emerged as technically superior to all the other forms of coating, as it isolates the steel from direct contact with the external corrosive environment while maintaining the other factors like bond strength and strength of steel, constant.

Galvanized coatings provide important advantages for the protection of reinforcement. Research and practical experience since the 1950s have established that the corrosion resistance of galvanized steel reinforcement is superior to uncoated steel and any other type of protective measure. The corrosion protection by the galvanized coating ensures that the design strength of concrete is maintained and the possibility of surface rust staining and eventual corrosion of reinforcement and spelling of concrete are removed.

Galvanising

Galvanising process is basically the process of formation of series of alloy layers formed as a result of a metallurgical reaction between steel and zinc.

The process of galvanisation consists of five distinct steps which are briefly described below:

- *Pickling*: This is the first step and involves the dipping on steel into an acidic solution containing hydrochloric acid to remove the surface scales, rust and other impurities.
- The second step includes rinsing of the steel to neutralize the effect of the acidic and obtain a clean surface.
- The steel is then dipped in the flux solution to prepare the surface to absorb the coating of zinc and also ensure good bonding between the steel and zinc.
- Drying is then undertaken to remove the unwanted moisture and increase the steel temperature to the required level for complete coating.

- The steel is then dipped into molten zinc at 450-460 C, and reaction is initiated between steel and zinc. This forms a series of inter metallic alloy layers called Eta, Zeta, Delta, Gamma having varying composition of zinc and iron.

Thus in the case of galvanising, the protector (zinc) and the protected (steel) become one, which distinguishes itself from other forms of coating which is merely a physical coating over the steel held to the surface by the adhesive forces only.

Also, these alloy layers impart the hardness and abrasion resistance to the galvanised rebar which prevents peeling off of the layers in handling, and storage unlike in other coatings. This protects the steel not only inside the concrete but also in handling and storage at site. The galvanising bath is typically 8 to 13 metres in length; also double dipping can be done for longer bars. Approved galvanizes are present across India who caters to the demand in individual geographical areas.

Properties of Galvanized Coating

- *Uniform coating thickness*: The coating thickness remains the same across the entire length of the bar. Even edges have the same coating thickness since it is a metallurgical bond and reaction proceeds from inside t outside.
- *Cathodic protection*: This is a unique property of zinc coating which covers up any holidays/scratches that may occur in the coating. This is due to the electrochemical property of zinc which is more electro positive than steel and has a tendency to donate electrons to exposed steel. Hence if a holiday occurs in the zinc coating the zinc from surrounding area comes and covers it up.
- *Possibility of bending of galvanised steel*: The rebars coated can be bent after galvanization without cracking/peeling up of coating. This is by virtue of the coating being a metallurgical coating and zinc having good ductility. Hence galvanized bars can be bend at the site itself.
- *Bond strength*: The results of extensive programmes of pullout testing by a number of researches reveal no significant difference in the bond strengths of black and galvanized steel deformed (*i.e.* ribbed) reinforcing bars in concrete.
- *Physical properties of steel*: Galvanizing is done at a temperature of 450-460 degrees centigrade for duration of 2

minutes. There is change in physical properties of steel beyond a temperature level of 700-750 degrees centigrade. Hence the researches have shown that there is no change in the physical characterstics of steel due to galvanising. This has been proven conclusively in all the studies carried out on galvanised rebars.

- *Welding*: All kinds of welding (including arc welding) can be performed as per the guidelines of American Welding Society. All types of welding (including arc welding) can be performed on galvanized rebar's as per CECRI and IIT Bombay.

Benefits

- Protection to the steel during storage and construction prior to placing the concrete.
- Diminished effect of variations in concrete quality. Safeguards against poor workmanship, especially misplacement of reinforcement, poor compaction and inadequate curing.
- Delayed initiation of corrosion and the onset of cracking.
- Reduced likelihood of surface staining.
- Increased structural life of concrete, particularly where chloride contamination is likely.

Specification

Galvanising of reinforcing steel should be specified to comply with Indian Standard IS: 12594-1988 "Hot dip zinc coating on structural steel bars for concrete reinforcement-Specification". This is in line with the international standards on the same viz. ASTM A 767. The coating weight is 610 gsm *i.e.* 80-90 microns for general construction and 915 gsm *i.e.* 110-120 microns for severe marine environment.

- In a process for preheating a reduction gas or air employed in producing or refining of iron or steel wherein said reduction gas or air is passed through a preheater and from said preheater through a pipe to a producing or refining furnace for producing or refining said iron or steel the improvement wherein said preheater and said pipe comprise steel shells lined with an acid resisting cement for protecting the steel from corrosion, said acid resisting cement having on its inner surface averted from said steel shell a refractory lining, said cement comprising:
  - Quartz sand and refractory material in the form of hollow spheres, said spheres being 10 - 40 vol.% of the total volume of the quartz sand and spheres,

- Silicate binder for the cement
- Hardener.

- Process of claim 1, wherein said hollow spheres are 22 to 28 vol.% of the total volume of quartz sand and hollow spheres.
- Process of claim 1, the hollow spheres comprising corundum.
- Process of claim 1, wherein said cement is sprayed onto said steel shell to result in a lining thickness of 5 to 15 mm.
- Process of claim 1, wherein the binder is at least one of sodium and potassium silicate.
- Process of claim 1, wherein the amount of binder is 2 to 7 vol.%, the amount of hardener is 0.5 to 6 vol.%, and the balance is essentially said quartz sand and spheres.
- Process of claim 4, wherein the amount of binder is 4 to 6 vol.%, the amount of hardener is 1 to 4 vol.%, and the balance is essentially said quartz sand and spheres.
- Process of claim 1, wherein said furnace is communicated by said pipes alternately with one of at least two of said preheaters, said preheaters being regeneration preheaters which are alternately heated by flue gas and are available after heating for switching over for said preheating of said air or reduction gas.
- Process of claim 1, wherein said furnace is a blast furnace communicated by said pipes alternately with one of two or three of said regeneration preheaters, the hot one of said preheaters preheating said air, whereas the other one or two are heated by said flue gas.

Description

The problem that the hot and aggressive corrosive Cowper gases and condensates produced during blast furnace blast heating cause a great deal of damage to steel by corrosion, exists in the steel industry in particular. The corrosion effect of the Cowper gases can be explained as follows: It has to be seen that the Cowper work regeneratively *i.e.* at least one of the air preheaters is heated by flue gas (spent gas) up to a temperature of about 1450° C and that this flue gas contains SO_2 and SO_3 and water vapour which may form sulfuric acid by condensation.

The chance for this condensation is high if the refractory lining is not sealed and so the corrosive parts of the gas have the chance to condense on the cold steel shell.

Another proposal for a protective measure against corrosion relates to spraying an acid-resisting cement onto the inner surfaces of the metal, particularly steel, in danger of corroding (Merkblatt Oberflaechentechnik HOECHST, Saurekitt, Sept. 1974, A 1187 I D). The acid-resisting cement proposed there as a protective measure against corrosion should be applied onto the steel surfaces in danger of corroding in the form of a coating without joints of for example 10 - 25 mm according to the type of fireclay spraying process. It is known that these acid-resisting cements contain quartz sand, silicate binder and hardener. The quartz sand may be present as pure SiO_2.

However, quartz sand in the natural state will be generally used. Normally, a sodium or potassium silicate binder is introduced as the binder. The dry acid-resisting cement contains about 2 - 7 vol.% silicate binder, the addition of hardener is co-ordinated to this amount and the remainder is essentially quartz sand. The hardener can be a condensed aluminum phosphate as disclosed in the German Auslegeschrift 1,571,485 and German Patent 1,252,835 of the Hoechst Company, in amount of 0.5 to 6 vol.%, preferably 1.0 to 4 vol.%.

The Inventory

Working from the last-mentioned proposal for protection against corrosion for hot, aggressive media, especially gaseous media, the object of the present invention is to develop a means of protection against corrosion, whih better absorbs the expansion and contraction loads of steel sheets and of the refractory materials on the basis of the above explained fluctuations in temperature. This problem is particularly serious in the blast preheater and the hot-air pipes thereof, as considerable fluctuations in temperature arise due to the technical course in this case of usage.

Consequently, temperatures of up to over 1450° C are reached in the cupola of the blast preheater during the heating up periods, after which a considerable drop in temperature occurs after reversal in the blast period due to the cold blast air being heated or in the case of stoppages. It has surprisingly been proved that the durability of the protective layer against corrosion, with the given purpose of usage and particularly in the case of great fluctuations in temperature, can be considerably improved if 10 - 40% vol. of the total amount of quartz is replaced by refractory materials in the form of hollow spheres. An acid-resisting cement is preferably used in which 15 to 35% vol. of the total amount of quartz, in particular 22 to 28% vol, is replaced by refractory materials in the form of hollow spheres. The amount of binder can be 2 to 7 vol.%, preferably 4 to 6 vol.%. The amount of hardener can be 0.5 to 6 vol.%, preferably 1 to 4 vol.%. The composition of the binder and hardener can be as is disclosed above for prior art cements. Refractory materials in the form of hollow spheres belong in another context to the prior art. German Auslegeschrift 1,086,614 describes a refractory, sintered bonding body which contains a layer composed of refractory material in the form of hollow spheres for heat insulation. It can be further seen from this publication that it has been known for a long time that granules in the form of hollow spheres composed of refractory material are produced. Thus, alumina, quartz, mixtures of alumina and quartz and similar refractory materials are mentioned as material for the production of the hollow spheres.

An acid-resisting cement is preferably used according to the invention in which the refractory materials present in the form of hollow spheres of corundum (hollow sphere corundum). It remains to bear in mind that the refractory material added in the form of hollow spheres corresponds to the volume amount which was removed from the known acid-resisting cement in the form of quartz sand. This means in other words that the thickness of the layer applied does not change, however, the structure does. The thickness of the layer amounts in practice to between 5 and 15 mm. Very good results have been achieved with layer thickness of up to 8 mm. The particular advantages of the subject of the invention is to be seen in that the proposed acid-resisting cement is more flexible to fluctuations in temperature and therefore the movements of the refractory material and the steel shell are better absorbed. It is evident that the layer does not only act as a protection against corrosion but as an insulating layer for the movements which are caused by fluctuations in temperature. The layer consequently is very flexible. It has moreover been shown that this layer is less sensitive to subsequent moisture (wetting).

On the contrary, the known acid-resisting cement becomes relatively moist after spraying so that according to the prior art technique it may result in the sprayed-on elements falling, rolling or sliding away during overhead work, *e.g.* in the cupola of the blast preheater. This subsequent wetting is known to the specialist by the term "ionisation effect" and occurs on a considerably lower scale with the subject of the application. The acid-resisting cement proposed according to the invention is preferably used for the steel shell of the blast heating apparatus of a blast furnace and for the hot-air pipe system appertaining thereto. An acid-resisting cement according to the invention was used as protective means against corrosion to protect the cupola and the upper portion of the shaft of the blast preheater. The said acid-resisting cement, comprising quartz sand, a sodium silicate binder, a usual hardener and hollow sphere corundum, was composed as follows: 70 vol.% quartz sand; 4 vol.% sodium silicate binder, 1 vol.% of hardener, and 25 vol.% as hollow sphere corundum. The diameter of the hollow sphere corundum used was between about 0.06 and 2 mm. The diameter of the greater part of the hollow sphere corundum was about 1mm.

The acid-resisting cement serving as a protective means against corrosion was applied by the spraying process to the surface of the steel without anchorage by means of a rotor machine, which works continuously according to a revolving system. The dry means is mixed shortly before spraying with 14 wt.% water, based on the total weight of the dry means. The surfaces which were to be protected were continuously covered with the protective means against corrosion without seams, the thickness of the layer amounting to about 8mm. The sprayed-on means was set after about 12 hours with the temperature of the sheet metal at about 20° C. Further lining comprising insulating layers and a refractory layer was then applied. Preferably the further lining consists of bricks. The described blast heating apparatus of blast furnace reaches a temperature of about 1450° - 1500° C. in the area of the cupola and the upper shaft during the heating period, whereas in the blast period the temperature in the described areas decreases to about 1050° C. Despite these difference in temperature and the pressure connected therewith, which lead to the gradual loosening of the refractory lining and consequently permits corrosive means to penetrate to the vicinity of the sheet metal shell, no damage has occured after usage for one year and after usage for 2 years.

With comparable positive result the blast air steel pipes belonging to the blast heating apparatus of the blast furance were sprayed with the protective means against corrosion according to the invention and then lined with two insulating layers and a refractory layer.

The invention has been explained in detail by the example blast furnace and regeneration air preheaters as for this example the fluctuations in temperature during the regenerative phases amount to about 500° C.. It has to be recognized that the invention is also suitable for protecting the steel shell if the preheated is used to preheat reduction gas as in necessary for alternative processes for the blast furnace process.

Chapter 3

Polymer Chemistry

Polymer chemistry or macromolecular chemistry is a multidisciplinary science that deals with the chemical synthesis and chemical properties of polymers or macromolecules. According to IUPAC recommendations, macromolecules refer to the individual molecular chains and are the domain of chemistry. Polymers describe the bulk properties of polymer materials and belong to the field of polymer physics as a subfield of physics.

Polymer chemistry is that branch of one,which deals with the study of synthesis and properties of macromolecules.

- Biopolymers produced by living organisms:
  - structural proteins: collagen, keratin, elastin...
  - chemically functional proteins: enzymes, hormones, transport proteins...
  - structural polysaccharides: cellulose, chitin...
  - storage polysaccharides: starch, glycogen...
  - nucleic acids: DNA, RNA.
- Synthetic polymers used for plastics—fibres, paints, building materials, furniture, mechanical parts, adhesives:
  - thermoplastics: polyethylene, Teflon polystyrene, polypropylene, polyester, polyurethane, polymethyl methacrylate, polyvinyl chloride, nylon, rayon, celluloid, silicone, fiberglass...
  - thermosetting plastics: vulcanized rubber, Bakelite, Kevlar, epoxy...

Polymers form by polymerization of monomers. A polymer is chemically described by its degree of polymerisation, molar mass distribution, tacticity, copolymer distribution, the degree of branching, by its end-groups, crosslinks, crystallinity and thermal properties such as its glass transition temperature and melting temperature. Polymers in solution have special characteristics with respect to solubility, viscosity and gelation.

History

The work of Henri Braconnot in 1777 and the work ofChristian Schönbein in 1846 led to the discovery of nitrocellulose, which, when treated with camphor produced celluloid. Dissolved in ether or acetone, it is collodion, used as a wound dressing since the U.S. Civil War. Cellulose acetate was first prepared in 1865. In 1834, Friedrich Ludersdorf and Nathaniel Hayward independently discovered that adding sulfur to raw natural rubber (polyisoprene) helped prevent the material from becoming sticky. In 1844 Charles Goodyear received a U.S. patent for vulcanizing rubber with sulfur and heat. Thomas Hancock had received a patent for the same process in the UK the year before.

In 1884 Hilaire de Chardonnet started the first artificial fiber plant based on regenerated cellulose, or viscose rayon, as a substitute for silk, but it was very flammable. In 1907 Leo Baekeland invented the first synthetic polymer, a thermosetting phenol-formaldehyde resin called Bakelite. Cellophane was invented in 1908 by Jocques Brandenberger who squirted sheets of viscose rayon into an acid bath.

In 1922 Hermann Staudinger was the first to propose that polymers consisted of long chains of atoms held together by covalent bonds. He also proposed to name these compounds macromolecules. Before that, scientists believed that polymers were clusters of small molecules (called colloids), without definite molecular weights, held together by an unknown force. Staudinger received the Nobel Prize in Chemistry in 1953.

Wallace Carothers invented the first synthetic rubber called neoprene in 1931, the first polyester, and went on to invent nylon, a true silk replacement, in 1935. Paul Flory was awarded the Nobel Prize in Chemistry in 1974 for his work on polymer random coil configurations in solution in the 1950s. Stephanie Kwolek developed an aramid, or aromatic nylon named Kevlar, patented in 1966.

There are now a large number of commercial polymers, including composite materials such as carbon fiber-epoxy, polystyrene-polybutadiene (HIPS), acrylonitrile-butadiene-styrene (ABS), and other such materials that combine the best properties of their various components, including polymers designed to work at high temperatures in automobile engines.

Polymer

A polymer is a large molecule (macromolecule) composed of repeating structural units typically connected by covalent chemical bonds. While *polymer* in popular usage suggests plastic, the term actually refers to a large class of natural and synthetic materials with a variety of properties. Due to the extraordinary range of properties accessible in polymeric materials, they have come to play an essential and ubiquitous role in everyday life- from plastics and elastomers on the one hand to natural biopolymers such as DNA and proteins that are essential for life on the other. A simple example is polyethylene, whose repeating unit is based on ethylene (IUPAC name *ethene*) monomer. Most commonly, as in this example, the continuously linked backbone of a polymer used for the preparation of plastics consists mainly of carbon atoms. However, other structures do exist; for example, elements such as silicon form familiar materials such as silicones, examples being silly putty and waterproof plumbing sealant. The backbone of DNA is in fact based on a phosphodiester bond, and repeating units of polysaccharides (e.g. cellulose) are joined together by glycosidic bonds via oxygen atoms.

Natural polymeric materials such as shellac, amber, and natural rubber have been in use for centuries. Biopolymers such as proteins and nucleic acids play crucial roles in biological processes. A variety of other natural polymers exist, such as cellulose, which is the main constituent of wood and paper.

The list of synthetic polymers includes synthetic rubber, Bakelite, neoprene, nylon, PVC, polystyrene, polyethylene, polypropylene, polyacrylonitrile, PVB, silicone, and many more.

Polymers are studied in the fields of polymer chemistry, polymer physics, and polymer science.

Etymology

The word *polymer* is derived from the Greek words *ðïëõ* (*poly*), meaning "many"; and *μÝñïò* (*meros*), meaning "part". The term was coined in 1833 by Jöns Jakob Berzelius, although his definition of a polymer was quite different from the modern definition.

Historical Development

Starting in 1811, Henri Braconnot did pioneering work in derivative cellulose compounds, perhaps the earliest important work in polymer science. The development of vulcanization later in the nineteenth century improved the durability of the natural polymer rubber, signifying the first popularized semi-synthetic polymer. In 1907, Leo Baekeland created the first completely synthetic polymer, Bakelite, by reacting phenol and formaldehyde at precisely controlled temperature and pressure. Bakelite was then publicly introduced in 1909.

Despite significant advances in synthesis and characterization of polymers, a correct understanding of polymer molecular structure did not emerge until the 1920s. Before then, scientists believed that polymers were clusters of small molecules (called colloids), without definite molecular weights, held together by an unknown force, a concept known as association theory. In 1922, Hermann Staudinger proposed that polymers consisted of long chains of atoms held together by covalent bonds, an idea which did not gain wide acceptance for over a decade and for which Staudinger was ultimately awarded the Nobel Prize. Work by Wallace Carothers in the 1920s also demonstrated that polymers could be synthesized rationally from their constituent monomers. An important contribution to synthetic polymer science was made by the Italian chemist Giulio Natta and the German chemist Karl Ziegler, who won the Nobel Prize in Chemistry in 1963 for the development of the Ziegler-Natta catalyst. Further recognition of the importance of polymers came with the award of the Nobel Prize in Chemistry in 1974 to Paul Flory, whose extensive work on polymers included the kinetics of step-growth polymerization and of addition polymerization, chain transfer, excluded volume, the Flory-Huggins solution theory, and the Flory convention.

Synthetic polymer materials such as nylon, polyethylene, Teflon, and silicone have formed the basis for a burgeoning polymer industry. These years have also shown significant developments in rational polymer synthesis. Most commercially important polymers today are entirely synthetic and produced in high volume on appropriately scaled organic synthetic techniques. Synthetic polymers today find application in nearly every industry and area of life. Polymers are widely used as adhesives and lubricants, as well as structural components for products ranging from children's toys to aircraft. They have been employed in a variety of biomedical applications ranging from implantable devices to controlled drug delivery. Polymers such as poly(methyl methacrylate) find application as photoresist materials used in semiconductor manufacturing and low-k dielectrics for use in high-performance microprocessors. Recently, polymers have also been employed as flexible substrates in the development of organic light-emitting diodes for electronic displays.

Polymer Synthesis

Polymerization is the process of combining many small molecules known as monomers into a covalently bonded chain. During the polymerization process, some chemical groups may be lost from each monomer. This is the case, for example, in the polymerization of PET polyester. The monomers are terephthalic acid ($HOOC-C_6H_4-COOH$) and ethylene glycol ($HO-CH_2-CH_2-OH$) but the repeating unit is $-OC-C_6H_4-COO-CH_2-CH_2-O-$, which corresponds to the combination of the two monomers with the loss of two water molecules. The distinct piece of each monomer that is incorporated into the polymer is known as a repeat unit or monomer residue.

Laboratory Synthesis

Laboratory synthetic methods are generally divided into two categories, step-growth polymerization and chain-growth polymerization. The essential difference between the two is that in chain growth polymerization, monomers are added to the chain one at a time only, whereas in step-growth polymerization chains of monomers may combine with one another directly. However, some newer methods such as plasma polymerization do not fit neatly into either category. Synthetic polymerization reactions may be carried out with or without a catalyst. Efforts towards rational synthesis of biopolymers via laboratory synthetic methods, especially artificial synthesis of proteins, is an area of intense research.

Biological Synthesis

There are three main classes of biopolymers: polysaccharides, polypeptides, and polynucleotides. In living cells, they may be synthesized by enzyme-mediated processes, such as the formation of DNA catalyzed by DNA polymerase. The synthesis of proteins involves multiple enzyme-mediated processes to transcribe genetic information from the DNA to RNA and subsequently translate that information to synthesize the specified protein from amino acids. The protein may be modified further following translation in order to provide appropriate structure and functioning.

Modification of Natural Polymers

Many commercially important polymers are synthesized by chemical modification of naturally occurring polymers. Prominent examples include the reaction of nitric acid and cellulose to form nitrocellulose and the formation of vulcanized rubber by heating natural rubber in the presence of sulphur.

Polymer Properties

Polymer properties are broadly divided into several classes based on the scale at which the property is defined as well as upon its physical basis. The most basic property of a polymer is the identity of its constituent monomers. A second set of properties, known as microstructure, essentially describe the arrangement of these monomers within the polymer at the scale of a single chain. These basic structural properties play a major role in determining bulk physical properties of the polymer, which describe how the polymer behaves as a continuous macroscopic material. Chemical properties, at the nano-scale, describe how the chains interact through various physical forces. At the macro-scale, they describe how the bulk polymer interacts with other chemicals and solvents.

Monomers/Repeat Units

The identity of the monomer residues (repeat units) comprising a polymer is its first and most important attribute. Polymer nomenclature is generally based upon the type of monomer residues comprising the polymer. Polymers that contain only a single type of repeat unit are known as homopolymers, while polymers containing a mixture of repeat units are known as copolymers. Poly(styrene), for example, is composed only of styrene monomer residues, and is therefore classified as a homopolymer. Ethylene-vinyl acetate, on the other hand, contains more than one variety of repeat unit and is thus a copolymer. Some biological polymers are composed of a variety of different but structurally related monomer residues; for example, polynucleotides such as DNA are composed of a variety of nucleotide subunits.

A polymer molecule containing ionizable subunits is known as a polyelectrolyte or ionomer.

Microstructure

The microstructure of a polymer (sometimes called configuration) relates to the physical arrangement of monomer residues along the backbone of the chain. These are the elements of polymer structure that require the breaking of a covalent bond in order to change. Structure has a strong influence on the other properties of a polymer. For example, two samples of natural rubber may exhibit different durability, even though their molecules comprise the same monomers.

An important microstructural feature determining polymer properties is the polymer architecture. The simplest polymer architecture is a linear chain: a single backbone with no branches. A related unbranching architecture is a ring polymer. A branched polymer molecule is composed of a main chain with one or more substituent side chains or branches. Special types of branched polymers include star polymers, comb polymers, brush polymers, ladders, and dendrimers.

Branching of polymer chains affects the ability of chains to slide past one another by altering intermolecular forces, in turn affecting bulk physical polymer properties. Long chain branches may increase polymer strength, toughness, and the glass transition temperature due to an increase in the number of entanglements per chain. The effect of such long-chain branches on the size of the polymer in solution is characterized by the branching index. Random length and atactic short chains, on the other hand, may reduce polymer strength due to disruption of organization and may likewise reduce the crystallinity of the polymer.

A good example of this effect is related to the range of physical attributes of polyethylene. High-density polyethylene (HDPE) has a very low degree of branching, is quite stiff, and is used in applications such as milk jugs. Low-density polyethylene (LDPE), on the other hand, has significant numbers of both long and short branches, is quite flexible.

Dendrimers are a special case of polymer where every monomer unit is branched. This tends to reduce intermolecular chain entanglement and crystallization. Alternatively, dendritic polymers are not perfectly branched but share similar properties to dendrimers due to their high degree of branching.

The architecture of the polymer is often physically determined by the functionality of the monomers from which it is formed. This property of a monomer is defined as the number of reaction sites at which may form chemical covalent bonds. The basic functionality required for forming even a linear chain is two bonding sites. Higher functionality yields branched or even crosslinked or networked polymer chains.

An effect related to branching is chemical crosslinking-the formation of covalent bonds between chains. Crosslinking tends to increase T_g and increase strength and toughness. Among other applications, this process is used to strengthen rubbers in a process known as vulcanization, which is based on crosslinking by sulphur. Car tires, for example, are highly crosslinked in order to reduce the leaking of air out of the tire and to toughen their durability. Eraser rubber, on the other hand, is not crosslinked to allow flaking of the rubber and prevent damage to the paper.

A cross-link suggests a branch point from which four or more distinct chains emanate. A polymer molecule with a high degree of crosslinking is referred to as a polymer network. Sufficiently high crosslink concentrations may lead to the formation of an infinite network, also known as a gel, in which networks of chains are of unlimited extent — essentially all chains have linked into one molecule.

Chain Length

The physical properties of a polymer are strongly dependent on the size or length of the polymer chain.. For example, as chain length is increased, melting and boiling temperatures increase quickly. Impact resistance also tends to increase with chain length, as does the viscosity, or resistance to flow, of the polymer in its melt state. Chain length is related to melt viscosity roughly as $1:10^{3.2}$, so that a tenfold increase in polymer chain length results in a viscosity increase of over 1000 times Increasing chain length furthermore tends to decrease chain mobility, increase strength and toughness, and increase the glass transition temperature (T_g). This is a result of the increase in chain interactions such as Van der Waals attractions and entanglements that come with increased chain length. These interactions tend to fix the individual chains more strongly in position and resist deformations and matrix breakup, both at higher stresses and higher temperatures.

A common means of expressing the length of a chain is the degree of polymerization, which quanitifies the number of monomers incorporated into the chain. As with other molecules, a polymer's size may also be expressed in terms of molecular weight. Since synthetic polymerization techniques typically yield a polymer product including a range of molecular weights, the weight is often expressed statistically to describe the distribution of chain lengths present in the same. Common examples are the number average molecular weight and weight average molecular weight. The ratio of these two values is the polydispersity index, commonly used to express the "width" of the molecular weight distribution. A final measurement is contour length, which can be understood as the length of the chain backbone in its fully extended state.

The flexibility of an unbranched chain polymer is characterized by its persistence length.

Monomer Arrangement in Copolymers

Monomers within a copolymer may be organized along the backbone in a variety of ways.

- Alternating copolymers possess regularly alternating monomer residues.
- Periodic copolymers have monomer residue types arranged in a repeating sequence..
- Statistical copolymers have monomer residues arranged according to a known statistical rule. A statistical copolymer in which the probability of finding a particular type of monomer residue at an particular point in the chain is independent of the types of surrounding monomer residue may be referred to as a truly random copolymer.
- Block copolymers have two or more homopolymer subunits linked by covalent bonds. Polymers with two or three blocks of two distinct chemical species (e.g., A and B) are called diblock copolymers and triblock copolymers, respectively. Polymers with three blocks, each of a different chemical species (e.g., A, B, and C) are termed triblock terpolymers.
- Graft or grafted copolymers contain side chains that have a different composition or configuration than the main chain.

Tacticity

Tacticity describes the relative stereochemistry of chiral centers in neighboring structural units within a macromolecule. There are three types: isotactic (all substituents on the same side), atactic (random placement of substituents), and syndiotactic (alternating placement of substituents).

Polymer Morphology

Polymer morphology generally describes the arrangement of chains in space and microscopic ordering of many polymer chains.

Crystallinity

When applied to polymers, the term *crystalline* has a somewhat ambiguous usage. In some cases, the term *crystalline* finds identical usage to that used in conventional crystallography. For example, the structure of a crystalline protein or polynucleotide, such as a sample prepared for x-ray crystallography, may be defined in terms of a conventional unit cell composed of one or more polymer molecules with cell dimensions of hundreds of angstroms or more.

A synthetic polymer may be lightly described as crystalline if it contains regions of three-dimensional ordering on atomic (rather than macromolecular) length scales, usually arising from intramolecular folding and/or stacking of adjacent chains. Synthetic polymers may consist of both crystalline and amorphous regions; the degree of crystallinity may be expressed in terms of a weight fraction or volume fraction of crystalline material. Few synthetic polymers are entirely crystalline.

The crystallinity of polymers is characterized by their degree of crystallinity, ranging from zero for a completely noncrystalline polymer to one for a theoretical completely crystalline polymer. Polymers with microcrystalline regions are generally tougher, more flexible and more impact-resistant than totally amorphous polymers.

Polymers with a degree of crystallinity approaching zero or one will tend to be transparent, while polymers with intermediate degrees of crystallinity will tend to be opaque due to light scattering by crystalline or glassy regions. Thus for many polymers, reduced crystallinity may also be associated with increased transparency.

Chain Conformation

The space occupied by a polymer molecule is generally expressed in terms of radius of gyration, which is an average distance from the centre of mass of the chain to the chain itself. Alternatively, it may be expressed in terms of pervaded volume, which is the volume of solution spanned by the polymer chain and scales with the cube of the radius of gyration.

Mechanical Properties

The bulk properties of a polymer are those most often of end-use interest. These are the properties that dictate how the polymer actually behaves on a macroscopic scale.

Tensile Strength

The tensile strength of a material quantifies how much stress the material will endure before failing. This is very important in applications that rely upon a polymer's physical strength or durability. For example, a rubber band with a higher tensile strength will hold a greater weight before snapping. In general tensile strength increases with polymer chain length and crosslinking of polymer chains.

Young's Modulus of Elasticity

Young's Modulus quantifies the elasticity of the polymer. It is defined, for small strains, as the ratio of rate of change of stress to strain. Like tensile strength, this is highly relevant in polymer applications involving the physical properties of polymers, such as rubber bands. The modulus is strongly dependent on temperature.

Transport Properties

Transport properties such as diffusivity relate to how rapidly molecules move through the polymer matrix. These are very important in many applications of polymers for films and membranes.

Melting Point

The term *melting point*, when applied to polymers, suggests not a solid-liquid phase transition but a transition from a crystalline or semi-crystalline phase to a solid amorphous phase. Though abbreviated as simply T_m, the property in question is more properly called the crystalline melting temperature. Among synthetic polymers, crystalline melting is only discussed with regards to thermoplastics, as thermosetting polymers will decompose at high temperatures rather than melt.

Boiling Point

The boiling point of a polymeric material is strongly dependent on chain length. High polymers with a large degree of polymerization do not exhibit a boiling point because they decompose before reaching theoretical boiling temperatures. For shorter oligomers, a boiling transition may be observed and will generally increase rapidly as chain length is increased.

Glass Transition Temperature

A parameter of particular interest in synthetic polymer manufacturing is the glass transition temperature (T_g), which describes the temperature at which amorphous polymers undergo a second-order phase transition from a rubbery, viscous amorphous solid, or from a crystalline solid (depending on the degree of crystallization) to a brittle, glassy amorphous solid. The glass transition temperature may be engineered by altering the degree of branching or crosslinking in the polymer or by the addition of plasticizer.

Mixing Behavior

In general, polymeric mixtures are far less miscible than mixtures of small molecule materials. This effect results from the fact that the driving force for mixing is usually entropy, not interaction energy. In other words, miscible materials usually form a solution not because their interaction with each other is more favorable than their self-interaction, but because of an increase in entropy and hence free energy associated with increasing the amount of volume available to each component. This increase in entropy scales with the number of particles (or moles) being mixed. Since polymeric molecules are much larger and hence generally have much higher specific volumes than small molecules, the number of molecules involved in a polymeric mixture are far less than the number in a small molecule mixture of equal volume.

The energetics of mixing, on the other hand, are comparable on a per volume basis for polymeric and small molecule mixtures. This tends to increase the free energy of mixing for polymer solutions and thus make solvation less favorable. Thus, concentrated solutions of polymers are far rarer than those of small molecules.

In dilute solution, the properties of the polymer are characterized by the interaction between the solvent and the polymer. In a good solvent, the polymer appears swollen and occupies a large volume.

In this scenario, intermolecular forces between the solvent and monomer subunits dominate over intramolecular interactions. In a bad solvent or poor solvent, intramolecular forces dominate and the chain contracts. In the theta solvent, or the state of the polymer solution where the value of the second virial coefficient becomes 0, the intermolecular polymer-solvent repulsion balances exactly the intramolecular monomer-monomer attraction. Under the theta condition (also called the Flory condition), the polymer behaves like an ideal random coil.

Inclusion of Plasticizers

Inclusion of plasticizers tends to lower T_g and increase polymer flexibility. Plasticizers are generally small molecules that are chemically similar to the polymer and create gaps between polymer chains for greater mobility and reduced interchain interactions. A good example of the action of plasticizers is related to polyvinylchlorides or PVCs. A uPVC, or unplasticized polyvinylchloride, is used for things such as pipes. A pipe has no plasticizers in it, because it needs to remain strong and heat-resistant. Plasticized PVC is used for clothing for a flexible quality. Plasticizers are also put in some types of cling film to make the polymer more flexible.

Chemical Properties

The attractive forces between polymer chains play a large part in determining a polymer's properties. Because polymer chains are so long, these interchain forces are amplified far beyond the attractions between conventional molecules. Different side groups on the polymer can lend the polymer to ionic bonding or hydrogen bonding between its own chains. These stronger forces typically result in higher tensile strength and higher crystalline melting points.

The intermolecular forces in polymers can be affected by dipoles in the monomer units. Polymers containing amide or carbonyl groups can form hydrogen bonds between adjacent chains; the partially positively charged hydrogen atoms in N-H groups of one chain are strongly attracted to the partially negatively charged oxygen atoms in C=O groups on another.

These strong hydrogen bonds, for example, result in the high tensile strength and melting point of polymers containing urethane or urea linkages. Polyesters have dipole-dipole bonding between the oxygen atoms in C=O groups and the hydrogen atoms in H-C groups. Dipole bonding is not as strong as hydrogen bonding, so a polyester's melting point and strength are lower than Kevlar's (Twaron), but polyesters have greater flexibility.

Ethene, however, has no permanent dipole. The attractive forces between polyethylene chains arise from weak van der Waals forces. Molecules can be thought of as being surrounded by a cloud of negative electrons.

As two polymer chains approach, their electron clouds repel one another. This has the effect of lowering the electron density on one side of a polymer chain, creating a slight positive dipole on this side. This charge is enough to attract the second polymer chain. Van der Waals forces are quite weak, however, so polyethene can have a lower melting temperature compared to other polymers.

Standardized Polymer Nomenclature

There are multiple conventions for naming polymer substances. Many commonly used polymers, such as those found in consumer products, are referred to by a common or trivial name. The trivial name is assigned based on historical precedent or popular usage rather than a standardized naming convention. Both the American Chemical Society and IUPAC have proposed standardized naming conventions; the ACS and IUPAC conventions are similar but not identical. Examples of the differences between the various naming conventions are given in the table below:

Common Name	*ACS Name*	*IUPAC Name*
Poly (ethylene oxide) or (PEO)	poly(oxyethylene)	poly(oxyethene)
Poly (ethylene terephthalate) or (PET)	poly(oxy-1,2-ethanediy- loxycarbonyl -1,4 phenylenecarbonyl)	poly (oxyetheneoxy- terephth= aloyl)
Nylon	poly[amino(1-oxo- 1,6-hexanediyl)]	poly[amino(1-oxohexan-1,6-diyl)]

In both standardized conventions, the polymers' names are intended to reflect the monomer(s) from which they are synthesized rather than the precise nature of the repeating subunit.

Polymer Characterization

The characterization of a polymer requires several parameters which need to be specified. This is because a polymer actually consists of a statistical distribution of chains of varying lengths, and each chain consists of monomer residues which affect its properties.

A variety of lab techniques are used to determine the properties of polymers. Techniques such as wide angle X-ray scattering, small angle X-ray scattering, and small angle neutron scattering are used to determine the crystalline structure of polymers. Gel permeation chromatography is used to determine the number average molecular weight, weight average molecular weight, and polydispersity. FTIR, Raman and NMR can be used to determine composition. Thermal properties such as the glass transition temperature and melting point can be determined by differential scanning calorimetry and dynamic mechanical analysis.

Pyrolysis followed by analysis of the fragments is one more technique for determining the possible structure of the polymer. Thermogravimetry is a useful technique to evaluate the thermal stability of the polymer. Detailed analyses of TG curves also allow us to know a bit of the phase segregation in polymers. Rheological properties are also commonly used to help determine molecular architecture (molecular weight, molecular weight distribution and branching)as well as to understand how the polymer will process, through measurements of the polymer in the melt phase.

Another Polymer characterization technique is Automatic Continuous Online Monitoring of Polymerization Reactions (ACOMP) which provides real-time characterization of polymerization reactions. It can be used as an analytical method in R&D, as a tool for reaction optimization at the bench and pilot plant level and, eventually, for feedback control of full-scale reactors.

ACOMP measures in a model-independent fashion the evolution of average molar mass and intrinsic viscosity, monomer conversion kinetics and, in the case of copolymers, also the average composition drift and distribution. It is applicable in the areas of free radical and controlled radical homo- and copolymerization, polyelectrolyte synthesis,heterogeneous phase reactions, including emulsion polymerization, adaptation to batch and continuous reactors, and modifications of polymers.

Polymer Degradation

A plastic item with thirty years of exposure to heat and cold, brake fluid, and sunlight. Notice the discoloration, swollen dimensions, and tiny splits running through the material

Polymer degradation is a change in the properties—tensile strength, colour, shape, etc.—of a polymer or polymer-based product under the influence of one or more environmental factors, such as heat, light, chemicals and, in some cases, galvanic action. It is often due to the hydrolysis of the bonds connecting the polymer chain, which in turn leads to a decrease in the molecular mass of the polymer. These changes may be undesirable, such as changes during use, or desirable, as in biodegradation or deliberately lowering the molecular mass of a polymer. Such changes occur primarily because of the effect of these factors on the chemical composition of the polymer. Ozone cracking and UV degradation are specific failure modes for certain polymers.

A recent finding is that polymer degradation may occur through galvanic action. In 1990, Michael Faudree discovered that imide-linked resins in CFRP (carbon fiber reinforced polymers) composites degrade when bare composite is coupled with an active metal in saline, i.e. salt water environments. Polymers affected include bismaleimides (BMI), condensation polyimides, triazines, and blends thereof. Degradation occurs in the form of dissolved resin and loose fibres. Hydroxyl ions are generated at the graphite cathode attacking the O-C-N bond in the polyimide structure. This phenomenon, that polymers can undergo galvanic corrosion like metals do has been referred to as the "Faudree Effect". Standard corrosion protection procedures were found to prevent polymer degradation under most conditions. The degradation of polymers to form smaller molecules may proceed by random scission or specific scission. The degradation of polyethylene occurs by random scission—a random breakage of the linkages (bonds) that hold the atoms of the polymer together. When heated above 450 °C it degrades to form a mixture of hydrocarbons. Other polymers—like polyalphamethylstyrene—undergo specific chain scission with breakage occurring only at the ends. They literally unzip or depolymerize to become the constituent monomer.

However, the degradation process can be useful from the viewpoints of understanding the structure of a polymer or recycling/ reusing the polymer waste to prevent or reduce environmental pollution. Polylactic acid and polyglycolic acid, for example, are two polymers that are useful for their ability to degrade under aqueous conditions.

A copolymer of these polymers is used for biomedical applications, such as hydrolysable stitches that degrade over time after they are applied to a wound. These materials can also be used for plastics that will degrade over time after they are used and will therefore not remain as litter.

The sorting of polymer waste for recycling purposes may be facilitated by the use of the Resin identification codes developed by the Society of the Plastics Industry to identify the type of plastic.

Product Failure

In a finished product, such a change is to be prevented or delayed. Failure of safety-critical polymer components can cause serious accidents, such as fire in the case of cracked and degraded polymer fuel lines. Chlorine-induced cracking of acetal resin plumbing joints and polybutylene pipes has caused many serious floods in domestic properties, especially in the USA in the 1990s.

Traces of chlorine in the water supply attacked vulnerable polymers in the plastic plumbing, a problem which occurs faster if any of the parts have been poorly extruded or injection moulded. Attack of the acetal joint occurred because of faulty moulding leading to cracking along the threads of the fitting, which are serious stress concentrations.

Polymer oxidation has caused accidents involving medical devices. One of the oldest known failure modes is ozone cracking caused by chain scission when ozone gas attacks susceptible elastomers such as natural rubber and nitrile rubber.

They possess double bonds in their repeat units which are cleaved during ozonolysis. Cracks in fuel lines can penetrate the bore of the tube and cause fuel leakage. If cracking occurs in the engine compartment, electric sparks can ignite the gasoline and can cause a serious fire.

Fuel lines can also be attacked by another form of degradation: hydrolysis. Nylon 6,6 is susceptible to acid hydrolysis, and in one accident, a fractured fuel line led to a spillage of diesel into the road. If diesel fuel leaks onto the road, accidents to following cars can be caused by the slippery nature of the deposit, which is like black ice.

Polymerization

An example of alkene polymerization, in which each Styrene monomer unit's double bond reforms as a single bond with another styrene monomer and forms polystyrene.

In polymer chemistry, polymerization is a process of reacting monomer molecules together in a chemical reaction to form three-dimensional networks or polymer chains. There are many forms of polymerization and different systems exist to categorize them.

In chemical compounds, polymerization occurs via a variety of reaction mechanisms which vary in complexity due to functional groups present in reacting compounds and their inherent steric effects explained by VSEPR Theory.

In more straightforward polymerization, alkenes, which are relatively stable due to σ bonding between carbon atoms form polymers through relatively simple radical reactions; conversely, more complex reactions such as those that involve substitution at the carbonyl atom require more complex synthesis due to the way in which reacting molecules polymerize.

As alkenes can be formed in somewhat straightforward reaction mechanisms, they form useful compounds such as polyethylene and polyvinyl chloride (PVC) when undergoing radical reactions, which are produced in high tonnages each year due to their usefulness in manufacturing processes of commercial products, such as piping, insulation and packaging. Polymers such as PVC are generally referred to as "homopolymers" as they consist of repeated long chains or structures of the same monomer unit, whereas polymers that consist of more than one molecule are referred to as "co-polymers".

Other monomer units, such as formaldehyde hydrates or simple aldehydes, are able to polymerize themselves at quite low temperatures (>-80°C) to form trimers; molecules consisting of 3 monomer units which can cyclize to form ring cyclic structures, or undergo further reactions to form tetramers, or 4 monomer-unit compounds.

Further compounds either being referred to as oligomers in smaller molecules. Generally, because formaldehyde is an exceptionally reactive electrophile it allows nucleophillic addition of hemiacetal intermediates, which are generally short lived and relatively unstable "mid stage" compounds which react with other molecules present to form more stable polymeric compounds.

Polymerization that is not sufficiently moderated and proceeds at a fast rate can be very hazardous. This phenomenon is known as Hazardous polymerization and can cause fires and explosions.

Step-growth

Step growth polymers are defined as polymers formed by the stepwise reaction between functional groups of monomers. Most step growth polymers are also classified as condensation polymers, but not all step growth polymers (like polyurethanes formed from isocyanate and alcohol bifunctional monomers) release condensates. Step growth polymers increase in molecular weight at a very slow rate at lower conversions and only reach moderately high molecular weights at very high conversion (*i.e.* >95%).

To alleviate inconsistencies in these naming methods, adjusted definitions for condensation and addition polymers have been developed. A condensation polymer is defined as a polymer that involves elimination of small molecules during its synthesis, or contains functional groups as part of its backbone chain, or its repeat unit does not contain all the atoms present in the hypothetical monomer to which it can be degraded.

Chain-growth

Chain-growth polymerization or addition involves the linking together of molecules incorporating double or triple chemical bonds. These unsaturated *monomers* (the identical molecules which make up the polymers) have extra internal bonds which are able to break and link up with other monomers to form the repeating chain. Addition polymerization is involved in the manufacture of polymers such as polyethylene, polypropylene and polyvinyl chloride (PVC). A special case of addition polymerization leads to living polymerization.

In the radical polymerization of ethylene, its pi bond is broken and these two electrons rearrange to create a new propagating centre like the one that attacked it.

The form this propagating centre takes depends on the specific type of addition mechanism. There are several mechanisms through which this can be initiated.

The free radical mechanism was one of the first methods to be used. Free radicals are very reactive atoms or molecules which have unpaired electrons. Taking the polymerization of ethylene as an example, the free radical mechanism can be divided in to three stages: chain initiation, chain propagation and chain termination.

Polymerization of Ethylene

Free radical addition polymerization of ethylene must take place at high temperatures and pressures, approximately 300°C and 2000 At. While most other free radical polymerizations do not require such extreme temperatures and pressures, they do tend to lack control. One effect of this lack of control is a high degree of branching. Also, as termination occurs randomly, when two chains collide, it is impossible to control the length of individual chains.

A newer method of polymerization similar to free radical, but allowing more control involves the Ziegler-Natta catalyst especially with respect to polymer branching.

Other forms of addition polymerization include cationic addition polymerization and anionic addition polymerization. While not used to a large extent in industry yet due to stringent reaction conditions such as lack of water and oxygen, these methods provide ways to polymerize some monomers that cannot be polymerized by free radical methods such as polypropylene. Cationic and anionic mechanisms are also more ideally suited for living polymerizations, although free radical living polymerizations have also been developed.

Copolymer

A heteropolymer or copolymer is a polymer derived from two (or more) monomeric species, as opposed to a homopolymer where only one monomer is used. Copolymerization refers to methods used to chemically synthesize a copolymer.

Commercially relevant copolymers include ABS plastic, SBR, Nitrile rubber, styrene-acrylonitrile, styrene-isoprene-styrene (SIS) and ethylene-vinyl acetate.

Types of Copolymers

Since a copolymer consists of at least two types of [constitutional unit]s (also structural units), copolymers can be classified based on how these units are arranged along the chain. These include:

- Alternating copolymers with regular alternating A and B units.
- Periodic copolymers with A and B units arranged in a repeating sequence (e.g. $(A\text{-}B\text{-}A\text{-}B\text{-}B\text{-}A\text{-}A\text{-}A\text{-}A\text{-}B\text{-}B\text{-}B)_n$)
- Statistical Copolymers are copolymers in which the sequence of monomer residues follows a statistical rule. If the probability of finding a given type monomer residue at a particular point in the chain is equal to the mole fraction of that monomer residue in the chain, then the polymer may be referred to as a truly random copolymer.
- Block copolymers comprise two or more homopolymer subunits linked by covalent bonds. The union of the homopolymer subunits may require an intermediate non-repeating subunit, known as a junction block. Block copolymers with two or three distinct blocks are called diblock copolymers and triblock copolymers, respectively.

Copolymers may also be described in terms of the existence of or arrangement of branches in the polymer structure. Linear copolymers consist of a single main chain whereas branched copolymers consist of a single main chain with one or more polymeric side chains.

Other special types of branched copolymers include star copolymers, brush copolymers, and comb copolymers.

A terpolymer is a copolymer consisting of three distinct monomers. The term is derived from *ter* (Latin), meaning thrice, and polymer.

Graft Copolymers

Graft copolymers are a special type of branched copolymer in which the side chains are structurally distinct from the main chain. The illustration depicts a special case where the main chain and side chains are composed of distinct homopolymers. However, the individual chains of a graft copolymer may be homopolymers or copolymers. Note that different copolymer sequencing is sufficient to define a structural difference, thus an A-B diblock copolymer with A-B alternating copolymer side chains is properly called a graft copolymer.

For example, suppose we perform a free-radical polymerization of styrene in the presence of polybutadiene, a synthetic rubber, which retains one reactive C=C double bond per residue. We get polystyrene chains growing out in either direction from some of the places where there were double bonds, with a one-carbon rearrangement. Or to look at it the other way around, the result is a polystyrene backbone with polybutadiene chains growing out of it in both directions. This is an interesting copolymer variant in that one of the ingredients was a polymer to begin with.

As with block copolymers, the quasi-composite product has properties of both "components". In the example cited, the rubbery chains absorb energy when the substance is hit, so it is much less brittle than ordinary polystyrene. The product is called high-impact polystyrene, or HIPS.

Block Copolymers

A special kind of copolymer is called a "block copolymer". Block copolymers are made up of blocks of different polymerized monomers. For example, PS-b-PMMA is short for polystyrene-b-poly (methyl methacrylate) and is made by first polymerizing styrene, and then subsequently polymerizing MMA from the reactive end of the polystyrene chains. This polymer is a "diblock copolymer" because it contains two different chemical blocks. You can also make triblocks, tetrablocks, multiblocks, etc. Diblock copolymers are made using living polymerization techniques, such as atom transfer free radical polymerization (ATRP), reversible addition fragmentation chain transfer (RAFT), ring-opening metathesis polymerization (ROMP), and living cationic or living anionic polymerizations. An emerging technique is chain shuttling polymerization.

Recent research in block copolymers suggests that they may be useful in creating self-constructing fabrics with potential utility in semiconductor arrays (for example, computer memory devices) by assembling fine details atop a structured base created using conventional microlithography methods.

Phase Separation

Block copolymers are interesting because they can "microphase separate" to form periodic nanostructures, as in the styrene-butadiene-styrene block copolymer shown at right. The polymer is known as Kraton and is used for shoe soles and adhesives. Owing to the microfine structure, the transmission electron microscope or TEM was needed to examine the structure.

The butadiene matrix was stained with osmium tetroxide to provide contrast in the image. The material was made by living polymerization so that the blocks are almost monodisperse, so helping to create a very regular microstructure. The molecular weight of the polystyrene blocks in the main picture is 102,000; the inset picture has a molecular weight of 91,000, producing slightly smaller domains.

Microphase separation is a situation similar to that of oil and water. Oil and water are immiscible-they phase separate. Due to incompatibility between the blocks, block copolymers undergo a similar phase separation. Because the blocks are covalently bonded to each other, they cannot demix macroscopically as water and oil.

In "microphase separation" the blocks form nanometer-sized structures. Depending on the relative lengths of each block, several morphologies can be obtained. In diblock copolymers, sufficiently different block lengths lead to nanometer-sized spheres of one block in a matrix of the second (for example PMMA in polystyrene).

By using less different block lengths, a hexagonally-packed-cylinder geometry can be obtained. Blocks of similar length form layers (often called lamellae in the technical literature). Between the cylindrical and lamellar phase is the gyroid phase. The nanoscale structures created from block copolymers could potentially be used for creating devices for use in computer memory, nanoscale-templating and nanoscale separations.

Polymer scientists use thermodynamics to describe how the different blocks interact. The product of the degree of polymerization, *N*, and the Flory-Huggins interaction parameter, ÷, gives an indication of how incompatible the two blocks are and whether or not they will microphase separate. For example, a diblock copolymer of symmetric composition will microphase separate if the product ÷*N* is greater than 10.5. If ÷*N* is less than 10.5, the blocks will mix and microphase separation is not observed.

Copolymer Equation

An alternating copolymer has the formula: -A-B-A-B-A-B-A-B-A-B-, or $-(\text{-A-B-})_n-$. The molar ratios of the monomer in the polymer is close to one, which happens when the reactivity ratios r_1 & r_2 are close to zero, as given by the Mayo-Lewis equation also called the copolymerization equation:

Where $r_1 = k_{11}/k_{12}$ & $r_2 = k_{22}/k_{21}$

Copolymer Engineering

Copolymerization is used to modify the properties of man-made plastics to specific needs, for example to reduce crystallinity, modify glass transition temperature or to improve solubility. It is a way of improving mechanical properties, in a technique known as rubber toughening. Elastomeric phases within a rigid matrix act as crack initiators, and so increase the energy absorption when the material is impacted for example. Acrylonitrile butadiene styrene is a common example.

Tacticity

Tacticity (from Greek 'taktikos': of or relating to arrangement or order) is the relative stereochemistry of adjacent chiral centers within a macromolecule. The practical significance of tacticity rests in the link between tacticity and the physical properties of the polymer. The regularity of the macromolecular structure influences the degree to which it has rigid, crystalline long range order or flexible, amorphous long range disorder. Precise knowledge of tacticity of a polymer also helps understanding at what temperature a polymer melts, how soluble it is in a solvent and its mechanical properties.

A tactic macromolecule in the IUPAC definition is a macromolecule in which essentially all the configurational (repeating) units are identical. Tacticity is particularly significant in vinyl polymers of the type $-H_2C-CH(R)-$ where each repeating unit with a substituent R on one side of the polymer backbone is followed by the next repeating unit with the substituent on the same side as the previous one, the other side as the previous one or positioned randomly with respect to the previous one. In a hydrocarbon macromolecule with all carbon atoms making up the backbone in a tetrahedral molecular geometry, the zigzag backbone is in the paper plane with the substituents either sticking out of the paper or retreating into the paper.

Diads

Two adjacent structural units in a polymer molecule constitute a diad. If the diad consists of two identically oriented units, the diad is called a meso diad reflecting similar features as a meso compound. If the diad consists of units oriented in opposition, the diad is called a racemo diad as in a racemic compound. In the case of vinyl polymer molecules, a meso diad is one in which the substituent carbon chains are oriented on the same side of the polymer backbone.

Triads

The stereochemistry of macromolecules can be defined even more precisely with the introduction of triads. An isotactic triad (mm) is made up of two adjacent meso diads, a syndiotactic triad (rr) consists of two adjacent racemo diads and a heterotactic triad (rm) is composed of a meso diad adjacent to a racemo diad. The mass fraction of isotactic (mm) triads is a common quantitative measure of tacticity.

When the stereochemistry of a macromolecule is considered to be a Bernoulli process, triad composition can be calculated from the probability of finding meso diads (P_m). When this probability is 0.25 then the probability of finding:

- an isotactic triad is P_m^2 or 0.0625
- an heterotactic triad is $2P_m(1\text{-}P_m)$ or 0.375
- a syndiotactic triad is $(1\text{-}P_m)^2$ or 0.5625

With a total probability of 1. Similar relationships with diads exist for tetrads.

Tetrads, Pentads, etc.

The definition of tetrads and pentads introduce further sophistication and precision to defining tacticity, especially when information on long-range ordering is desirable. Tacticity measurements obtained by Carbon-13 NMR are typically expressed in terms of the relative abundance of various pentads within the polymer molecule, e.g. *mmmm*, *mrrm*.

Other Conventions for Quantifying Tacticity

The primary convention for expressing tacticity is in terms of the relative weight fraction of triad or higher-order components, as described above. An alternative expression for tacticity is the average length of *meso* and *racemo* sequences within the polymer molecule. The average meso sequence length may be approximated from the relative abundance of pentads as follows:

Polymers

Isotactic Polymers

Isotactic polymers are composed of isotactic macromolecules (IUPAC definition). In isotactic macromolecules all the substituents are located on the same side of the macromolecular backbone. An isotactic macromolecule consists of 100% meso diads. Polypropylene formed by Ziegler-Natta catalysis is an isotactic polymer. Isotactic polymers are usually semicrystalline and often form a helix configuration.

Syndiotactic Polymers

In syndiotactic or syntactic macromolecules the substituents have alternate positions along the chain. The macromolecule consists 100% of racemo diads. Syndiotactic polystyrene, made by metallocene catalysis polymerisation, is crystalline with a melting point of 161 °C.

Atactic Polymers

In atactic macromolecules the substituents are placed randomly along the chain. The percentage of meso diads is between 1 and 99%. With the aid of spectroscopic techniques such as NMR it is possible to pinpoint the composition of a polymer in terms of the percentages for each triad.

Polymers that are formed by free-radical mechanisms such as polyvinylchloride are usually atactic. Due to their random nature atactic polymers are usually amorphous. In hemiisotactic macromolecules every other repeat unit has a random substituent.

Atactic polymers are technologically very important. A good example is polystyrene (PS). If a special catalyst is used in its synthesis it is possible to obtain the syndiotactic version of this polymer, but most industrial polystyrene produced is atactic. The two materials have very different properties because the irregular structure of the atactic version makes it impossible for the polymer chains to stack in a regular fashion. The result is that, whereas syndiotactic PS is a semicrystalline material, the more common atactic version cannot crystallize and forms a *glass* instead. This example is quite general in that many polymers of economic importance are atactic glass formers.

Head/tail Configuration

In vinyl polymers the complete configuration can be further described by defining polymer head/tail configuration. In a regular macromolecule all monomer units are normally linked in a head to tail configuration so that all â-substituents are separated by three carbon atoms. In head to head configuration this separation is only by 2 carbon atoms and the separation with tail to tail configuration is by 4 atoms. Head/tail configurations are not part of polymer tacticity but should be taken into account when considering polymer defects.

Techniques for Measuring Tacticity

Tacticity may be measured directly using proton or carbon-13 NMR. This technique enables quantification of the tacticity distribution by comparison of peak areas or integral ranges correspinding to known diads (r, m), triads (mm, rm+mr, rr) and/or higher order n-ads depending on spectral resolution. In cases of limited resolution stochastic methods such as Bernoullian or Markovian analysis may also be used to fit the distribution and back then forward predict higher nads and calculate the isotacticity of the polymer to the desired level.

Other techniques sensitive to tacticity include x-ray powder diffraction, secondary ion mass spectroscopy (SIMS), vibrational spectroscopy (FTIR) and especially two-dimensional techniques. Tacticity may also be inferred by measuring another physical property, such as melting temperature, when the relationship between tacticity and that property is well-established.

Chapter 4

Polymerisation

The field of polymer photochemistry continues to be an active area in applied photochemistry, with many topics having strong academic and industrial interest. Photopolymerization and photocuring science and technologies, as before, continue to be developed, particularly with regard towards designing novel and specific initiators and materials for specialist applications. Interest in active ionic initiators and radical-ionic processes continues to expand, while the photocrosslinking of polymers is attractive in terms of enhancing the physical and mechanical properties of electronic materials and the development of liquid crystalline materials.

The optical properties of polymers continues to be an active area of strong development, with a continued growth in photochromic and liquid crystalline materials. Last year saw a major literature explosion in LED's (light emitting diodes). The photooxidation of polymers on the other hand continues to decline at a low profile. Bio- and photodegradable plastics are important for agricultural usage although academic interest here is again minimal. The same applies to polymer stabilization, where commercial applications dominate very much, with emphasis on the practical use of stabilizers. For dyes and pigments stability is a major issue, especially in catalytic chemistry.

Photopolymerization

A number of reviews of topical and commercial interest have appeared on the subject of radiation curing science and technology. There continues to be much development in photoinitiators, with articles concentrating on organometallics and anion generators, metallocenes, polymeric systems and laser and general photoinitiation. The influence of heteroatoms in the photocuring of multifunctional monomers and prepolymers has been assessed and reviewed, with the sulfide group having a significant influence over that of the ether functionality. New lamp technologies are on the scene as are curable paints with improved drying times.

New environmentally friendly coatings are also available for UV curing on furniture, while adhesives continue to expand in the UV arena, with epoxy and cyanoacrylate resins having specialized interest. Epoxy resins are also highlighted for their non-gas generation during curing, while the modeling of photocuring reactions continues to expand our understanding of reaction rates and gel effects., The water resistance of emulsions has important implications, as do photosensitive polyimides for electronic circuitry boards. Sulfur compounds continue to attract interest as photoinitiators, and these have been reviewed in detail.

Photoinitiated Addition Polymerization

Photoinitiators, and their performance and mode of action, continue to attract interest. DTCS (*N,N*-diethy laminodithio-carbamoylmethylstyrene) has been found to play an important role as an inimer capable of producing living radicals for vinyl polymerizations. The bifunctional character of *N,N*-dimethyl-4-nitroaniline enables it to operate as an exceptional initiator. The presence of electron donor-acceptor groups allows effective electron transfer to take place as well as offering excitation energy transfer to other tertiary amine molecules such as *N,N*-dimethylaniline. In such cases the latter undergoes rapid consumption in comparison with the former, *i.e.,* at a rate of 165:1. Sulfur compounds such as morpholinyl-2-benzothiazole disulfide are active initiators for polymerizing acrylic monomers. This compound undergoes significant degradative initiator transfer as does diphenyl diselenide. Sulfur dioxide, on the other hand, forms charge-transfer complexes with piperazine and *N*-acrylphenothiazine., The ratios of the complexes were important, with the piperazine locking into the PMMA chains. Benzophenone with ketosulfone moieties undergo a b-cleavage reaction, while in the presence of an amine it also produces a ketyl radical.

The nature of the excited states are important in this reaction, and this was also the conclusion of other work on novel bifunctional benzophenones. In this regard a two-photon absorption process has been considered for the effective polymerization of deep layers of monomers. Here the use of near IR is believed to activate deeper reactions.

Although the authors stipulate that no absorption occurs in the near IR region, they nevertheless propose a second absorption process to account for the improved rate. In fact, the process is unlikely to be due to electronic absorption but rather to heat generation through IR absorption. Several coumarins have been found to be effective photoinitiators by undergoing electron transfer with other co-initiators such as oxime esters or titanocene, while coumarin-nitroxide has been found to be an excellent radical trap to give a strongly fluorescent alkoxylamine product. In this case the fluorescence buildup during polymerization is a direct measure of radical generation. Photoinduced radical generation has been confirmed by subphthalocyanines using laser flash photolysis, while a new process involving encapsulation of pyrilium salts has been developed for the controlled release of initiators during polymerization.

From the point-of-view of comparisons, the thioxanthone initiator (ITX) has been found to be much less effective than benzophenone in photoinitiation activity, while for azoisobutyronitrile its diffusion rate has been found to control the rate of polymerization. Metal based complexes have been found to be of value in the photopolymerization of aniline. The entrapment of ruthenium bipyridyl in a hectorite clay with*N*-phenyl-*p*-phenylenediamine is one such system, which forms a strongly absorbing CT template complex in the interlayers capable of growing polyaniline. Ru has also been used to catalyse the addition of aromatic ketones to vinyl groups for effective polymerization. On the other hand, a complex based on azidopentaaminecobalt (III) undergoes a photoredox mechanism to give free radicals, while monomer complexes of a phosphorus bridged ferrocenophane undergo ring-opening polymerization reactions.

A novel cyanine-dye-borate complex has been found to be an excellent IR sensitizer while eosin-Y has been found to be a powerful photoinitiator in the presence of *N*-methyldiethanolamine. In the latter case, however, dye bleaching was effective only in the presence of a third iodonium salt component. Visible light photosensitizers have also been developed based on azo-dye bound natural polymers such as gelatin. Here the binding of the dye to the polymer played a crucial role in the rate process. In the presence of riboflavin the photopolymerization rate of 2-hydroxyethyl methacrylate has been found to be dependent upon the structure of the co-amine synergist. The type of alkylamino radical was considered to be important.

Rose Bengal also forms a 1:2 complex with the ferrocinium salt, which exists mostly as a contact ion-pair. This complex is a highly efficient initiator and has been found to be more effective than the benzoyl phosphine oxides. The phenoxazine dye, resazurin, is also highly effective in the presence of amine co-initiators for the photopolymerization of acrylamide, while methylene blue works well with *N*-methyldiethanolamine and an iodonium salt.

The 2-isopropylthioxanthone derivative is also a powerful sensitizer in the presence of ketosulfone derivatives of benzophenone. Diblock copolymers of MMA and VAc (vinyl acetate) have been made using benzophenone as a photointiator, while MMA itself has been unusally photopolymerized with a complex of titanium dioxide and sodium sulfite. Phenolic molecules are shown to reduce not only the rate of photopolymerization of acrylic monomers but also their molecular weight through effective chain termination. Poly(sodium styrenesulfonate-block-*N*-vinylcarbazole) has been successfully prepared through nitroxide mediated living polymerization. Here camphor-sulfonic acid and acetic anhydride were used as rate accelerating additives. Similar studies have been undertaken by other workers on polystyrene end-tagged with initator groups. PMMA photopolymerized in the presence of bis(cyclopentadienyl)titanium dichloride and sparteine, as a chelating agent, has been found to be insoluble in most common solvents without any evidence of crosslinking. Other acrylates gave a similar effect, producing polymers of high thermal stability.

Other studies of interest include the photopolymerization of methacrylate monomers in a polystyrene matrix, acyl- phosphine oxide photoinitiators, formation of polyphenols through polymerization of 4-hydroxybenzaldehyde dimet- hylnitrone, quenching of imidazolyl radicals, poly (methacrylic acid) with camphorquinone moieties and radical transfer using an acetophenone/Fe tris(*N*,*N*-di-Et-dithioc-arbamate) system.

Cationic photoinitiators continue to grow in interest particularly with regard to iodonium salts. Diaryliodonium salts are photosensitized by thioxanthone and anthraquinone molecules in the visible spectrum, while triarylsulfonium salts. Phenothiazine compounds exhibit a similar behaviour, while a number of cationically polymerizable monomers have been made based on enol ether groups, epoxy functional inorganic hybrids, benzyl ether groups and carbazole monomers with vinyl-vinyl ether groups. In this regard, aryl free radicals can also be used to accelerate the rate of cationic polymerization by onium salts.

Other cationic initiators include allylic salts, 2-benzyl-2-(*N*,*N*dimethyl- 2-ethoxycarbonyl-1-propenyl) ammonium hexafluoroantimonate-1-(4- morpholinophenyl)-butane-1-one and *N*-

(2-ethoxycarbonyl-1-propenyl) ammonium hexaf- luoroantimonate) benzophenone for the photopoly-merization of cyclohexene oxide. Oxetanes have been investigated where the rate of initiation of polymerization is high, such as with 2-phenyloxetanes. Rates for monomers with ρ-methoxyphenyl groups were also high. A new photoinitiator, diaryliodonium hexafluoro-phosphate, has been developed, while in the anthracene sensitized photopolymerization of diphenyliodonium salts 9-phenyl-anthracene was found to be the main photoproduct.

By use of laser flash photolysis, anthracene in its lowest excited singlet state undergoes electron transfer to the diphenyliodonium cation during the initial step of photoacid generation. This process is then followed by fast chemical reactions generating 9-phenylanthracene and protons, and then back electron transfer from the diphenyliodine radical to the anthracene radical cation. The same salts have also been successfully used to photopolymerize 3 methyltetrahydrofuran, while polymers have also been made through the photolysis of susbstituted aryl cyclic sulfonium zwitterionic salts and 2-cyclopropyl-4-methylene-1,3-dioxolanes.

Complexes have been formed between phenothiazine and triphenylsulfonium salts, whereas 4-hexyloxysubstituted diazonium salts are highly effective initiators but with high oxygen sensitivity. Other cationic systems include 3-acrylamide-9-ethylcarbazole with diphenyliodonium salts, 2-oxalines, fluoroalkyl propenyl ethers, ((9-carbazolyl) methyl)thiirane with diaryliodonium salts and iron arene complexes and (9-oxo-9*H*-fluorene-2-yl)phenyliodonium hexafluoroantimonate (V) for epoxides. *N*-Substituted maleimides have been used to initiate the photo- polymerization of mixtures of maleimide and vinyl ether acrylates.

Maleimides and tertiary amines are also effective initiators for Type II polymerization of acrylates, as are cationic initiators for polyfunctional 1-propenyl ethers and 2,3-dihydrofuran. For maleimide-donor photoco-polymerizations, the total electron density, as well as the distribution within the carbon-carbon double bond in the donor monomer, influences the preference and ratio of reactivity. Furthermore, the rate of co-polymerization will also be influenced by the donor structure.

Globular latex particles have been made by photopolymerizing MMA and 2,3-diphenylbutadiene in poly(*N*-isopropylacrylamide) (PNIPAM). The latter template undergoes coil-to-globe transitions owing to co-nonsolvency effects, where the polymerization takes advantage of the hydrophobic nanodomains formed by collapsed PNIPAM chains in the solvent system. Apparently, only very low levels of the template polymer were found in the copolymer latex. In a similar fashion, the photopolymerization kinetics of acrylamide have been correlated with monomer organization in various phases of an LLC template. Here the polymerization rate is enhanced in more ordered LLC phases owing to orientation of the monomer molecules.

The propagation rate of MMA has also been measured in miniemulsions. In this case the molecular weight exhibits spikes or peaks due to pulses in the initiation process where compartmentalization takes place. Silica functionalized with azo groups has been used to produce bi- and tri-phase poly(styrene)-poly(butadiene), while electrically conductive polymers have been made by photopolymerizing poly(pyrrole) with Group 1B metal particles. Metallic nanostructures were identified in the matrix. In a similar fashion electrically conducting poly(pyrrole)-poly(vinyl alcohol) composities have been made by photopolymerizing with iron-arene salts. Methanol was found to influence the morphology of the matrix and its conductivity. Conductive poly(aniline) has also been prepared by photo-polymerizing aniline in the presence of Ru(II) bipyridyl, and methylviologen complexes with *N* phenyl-ρ-phenylenediamine, bound into Flemion film.

Ladder polymers of ρ-xylylenediammonium sorbate have been prepared by topochemical polymerization, while alkadiyne mono-layers have been formed on graphite surfaces. Other studies of interest include supramolecular structures of bis(methacryloyl oxyethylenecarbonyl-oxyethylene) oxy oligoetheracrylate, synthesis of chloromethyl derivatives with phenylacryloyl groups and photo- coplymerization of 9-(2,3-epoxypropyl) carbazole with THF.

Polyaniline is apparently prepared more efficiently under light than thermally, while vinylester resins with hydro- xymethyl groups exhibit superior properties after polymerization. Photopolymerization of carbon disulfide has been shown to produce nanoscale gold coloured polythiene films, while super adsorbent polymers based on poly(acrylic acid-sodium acrylate) are more effectively made via photopolymerization., Branched polystyrene has been successfully prepared from the photopolymerization of ρ-(phenylselenomethyl) styrene, and good membranes made from the photopolymerization of sodium 4-styrenesulfonate-divinylbenzene. Concentration studies on acrylamide for holographic recordings have shown that an increase in the monomer reduces scattering, but at the disadvantage of a slower reaction rate.

The same monomer, when copolymerized with phthalic anhydride, gave different coloured products depending upon the temperature, and also gave nanostructured gels with 2-hydroxyethyl methacrylate. The microhardness of C70 crystals can be enhanced through irradiation.

Photoiniferters have received some attention, with a new polymerizable system based on b-methacylyloxyethyl-2-*N*,*N*-diethyldithiocarbamyl acetate being developed for producing living polymers. Branched polystyrenes have also been prepared via the inimer *N*,*N*-diethylaminodithio-carbamoylmethylstyrene while end functional polystyrenes have been made with phenylseleno groups. In this way graft copolymers of PMMA and branched poly(styrene) can be made.

The electro-optical behaviour of copolymers of 2-hydroxyethyl acrylate and styrene has been examined polymerizations. With respect to bleachable photoinitiators, an unsteady one-dimensional model has been developed, accounting for initiator consumption and optical attenuation, and giving rise to relationships for the spatial and temporal variation of the local initiator concentration and its rate. The maximum rates of photoinitiated polymerization of monomers capable of hydrogen bonding have been measured by RT-FTIR. These monomers were found to exhibit higher rates than those which do not form hydrogen bonds.

A two-photon induced polymerization has been demonstrated at 710 nm using a thioxanthone and diarylidonium salt in an epoxide resin while fractal analysis has been applied to STM images of coniferyl alcohol. Polymerization frontals have been monitored via photobleachable initiators, and Raman scattering has been applied to C_{60} single crystals. At high laser intensities some thermal decomposition of the polymer occurs during photopolymerization. Well defined block copolymers of PMMA and polyalkyl acrylates have been synthesized with polynuclear aromatics at the junctions, while steady-state fluorescence measurements have been used to monitor the free radical polymerization of 2-vinylnaphthalene. Here the critical temperature of polymerization was found to be 60°C.

Photocrosslinking

Several initiator systems have been examined for photocuring reactions. These include cyanine dye-borate complexes for TMPTA, polyimides with built-in fullerene, safranine dye for styrenesulfonate,

N,N-dimethyl-*N*-(2-hydroxy-3-phenoxypropyl)amine-4-nitrobenzamide for polyethylene glycol diglycidyl ether-pentaerythritol tetrakis (mercaptoacetate) copolymer, quaternary ammonium salts with *N,N*-dimethldithiocarbamate anions for poly(glycidyl methacrylate) films, thioxanthones for 1,6-hexanediol diacrylate, oligomeic amines as co-synergists for triacrylates, self-curing mixtures of 1,2,4-trimellitic anhydride, epichlorohydrin and glycidy methacrylate, carbohydrates bound to a-hydroxyalkylphenones, camphorquinone for dimethacrylates, photoacid generating copolymers of *o*-nitrobenzyloxycarbonyl and iminiosulfonate groups, nonaqueous dispersions of particles with photolable a-heptadecylphenacyl ester stabilizer chains, 4-vinyl benzoyldiphenylphosphine oxide for hydro-gels, quaternary ammonium salts for phenolic resins, and poly(glycidyl methacrylate), iron-arene complexes for epoxides, heptamethine cyanine dyes for coatings, acetophenones for isocyanates and water-based acrylated polyesters.

A series of sytrene oxides have been identified as novel curable systems while a series of poly(cinnam-4'-ylmethyl methacrylate) derivatives undergo linear photoreactions with polarized light, offering useful LC applications. Other systems include the development of photocatalytic coatings with titanium dioxide pigments. Numerous studies have appeared on cationic photocuring. Trimethylolpropane tripropenyl ethers have been used as reference products for polyfunctional propenyl ethers and display reactivities comparable with those of vinyl ether resins using onium salts as initiators. Temperature has been to shown to have a profound effect on the cationic curing rate of 3,4-epoxycyclohexylmethyl-3,4-epoxycyclohexanecarboxylate, while 3,4-epoxy-1-butene has been found to undergo fast cure owing to the stabilization of the growing cationic chain end by the double bond formed during ring opening.

The compound 3-ethyl-3-(2- ethylhexyloxymethyl) oxetane is useful as a UV curable adhesive, while other work has shown that UV curable epoxy acrylate adhesives cure faster under nitrogen than in oxygen, indicating that radical processes possibly play a role in the cure.

Oxetanyl-functional silasesquioxanes give pollution free tough coatings, as do resins based on multifunctional acrylates with alkoxysilyl thiols and silicone-epoxies. Several calixarene derivatives with protective trimethylsilyl and other cyclic ether groups have been prepared, as have several bifunctional monomers with 2,3-dihydrofuran moieties via the Heck reaction. The latter exhibit very high reactivity during photocationic curing.

Alkyl oxetanes possessing excellent compatibility with other monomers have been developed as have a series of epoxides with functionalities that stabilize radicals, such as benzyl and allyl groups. These stable radicals can apparently induce redox reactions in onium salts, accelerating their decomposition. The addition of an epoxidized soya-bean oil to an aromatic epoxide apparently accelerates the cationic cure rate via a triarylsulfonium salt.

Cycloaliphatic diepoxides exhibit the same behaviour, giving a highly insoluble polymer network. The ring-opening of terminal epoxides results in the formation of macrocyclic oligoethers, giving in this case soft segments. Using mixed acrylic and epoxy systems together with free radical and cationic initiators, detailed cure anaylsis showed that the former cure more rapidly. Also, while oxygen was found not to have an affect on the cure rate, cationic cure was influenced by increasing humidity.

Diazo resins of poly(phosphoric acid) undergo rapid photoconversion from an ionically soluble into a covalently insoluble system. Other studies of interest include the photocrosslinking of poly(vinylphenol) with epoxy moieties, silane-epoxies, epoxy adhesives, negative working poly(etherimides) and hybrid curing of acrylic-epoxies.

Photocycloaddition reactions have shown that dialkyl (*Z*, *Z*)-muconates undergo three reactions, involving stereo- regular polymer formation, isomerism from *Z*, *Z* to *E*, *E* forms, followedby amorphous polymer formation. Liquid crystalline poly(aryl cinnamate) ketocoumarins have been investigated as effective initiators for the 2-2 cycloaddition reaction. Triplet sensitization results in disruption of the nematic mesophase and UV-visible hyperchromism.

For dimethyl 1,4-naphthalenediacrylate only poly- crystalline forms undergo cycloaddition reactions, while a new polymer of 5-(6-hexanediol cinnamate)-L-glutamate has been synthesized. Fluorine containing distyrylbenzene polymers have been shown to undergo both photoisomerism and 2 + 2 cycloaddition reactions. The presence of the fluorine groups accelerates the reactions. Studies on *p*-phenylenediacrylic acids have shown that while *p*-mondecyl esters only give oligomers, the monoethyl esters with an amide bond undergo photodimerization.

Mixed monolayers of liquid crystals of hydrophilic cyanobiphenyl with long alkyl chains and a poly(amic acid) alkylamine salt undergo phase separation upon irradiation, with degradation of the monolayer of poly(amic acid). The morphology of mixtures of a liquid crystal and a poly (ethylene glycol diacrylate) gave an 'islands-sea' structure on polymerization. A mechanism was developed to account for the various types of organizations, depending upon the initiation of the photocuring reaction.

LC poly(ester imides) with cinnamoyl groups form Grandjean textures on annealing, with a blue iridescent colour, which is frozen in after curing. Diethyl (*Z*, *Z*)-muconate exhibits a topotactic polymerization on irradiation. At low temperatures below 45°C the reaction is inhibited owing to small changes in packing alignments. Azimuthal anchoring energy coefficients have been measured or poly(vinyl cinnamate), while glass-forming nematic twin molecules have been made. Photoinduced diffusion during photocuring of a chiral-nematic phase yields a cholesteric network in which the helical pitch gradually changes over the cross-section of the film matrix.

For bifunctional symmetric and unsymmetric diepoxides with flexible alkylene chains, the phase transition form the crystalline to the liquid crystalline state could be varied with temperature. A polymer network with a liquid crystalline superstructure is formed if the polymerization is carried out in the LC phase. Other studies of interest include ferroelectric copolymers, a hybrid 3-(trimethoxysilyl)propyl methacrylate and a quaternary ammonium surfactant with a polymerizable methacrylate group, a semiflexible polyester with twin-spaced *p*-oxybenzoyl diads and an LC poly(methyl acrylate) contaiing both a mesogen biphenyl unit and a styryl-2-pyridine unit as a photosensitizer. Various methodologies in curing processes have been documented.

A rotating reaction-surface polymerization reactor has been developed for measuring stepgrowth and chain-growth polymerizations. High conversions and fast cures were obtained, giving polymers with superior properties. Surfaces for bonding have been modified by treating with acrylated initiators, while photolithographic processes have been modified through the use of composities. Alternative surfaces have been developed with photoreactive groups, and dark coloured coatings with carbon black have been cured effectively. Other pigmented systems have been effectively photocured using mixtures of a xanthene dye and a ferrocinium salt.

Effective photocuring systems have also been developed for filament winding, while shrinkage has been measured by TMA. Microlenses have been developed from polysilanes, while other workers have devised processes for developing lenses for telecommunications.

A manufacturing process has also been developed for composite carbon-fibre reinforced parts. Epoxy-polyamide formulations were examined in pre-pegs and a model was developed for layered curing. Although cure was affected, the properties of the final material were below the specifications for high performance parts. Optimization of dental resins for cure have been investigated, and the ionic conductivity of methacrylates has been found to increase to amaximum on doping with lithium perchlorate.

A novel oligo(poly(ethylene glycol)fumarate) macromer has been developed for functionalizing networks. By locking in to polypeptides, the cured system can be used for developing cell-adhesion specific sequences. Other studies include kinetic evaluations on 3D acrylates, ionic conductance of methacrylates doped with magnesium salts, laser simulated polymerizations, degradable diacrylates, use of high intensity irradiance for reducing initiator concentrations and electrooptical properties of cured films.

Monitoring methodologies for cure continue to be investigated. Isothermal microcalorimetry has been further investigated for real-time measurements of rates of cure for a number of resin systems. Structure and functionality are important in this regard.

Real-time FTIR and infrared analysis have also been utilized in many ways. These include studies and assessment of surface gradients, conversion rates for dye-borate cationics, cycloaddition of bifunctional epoxies with dimethacrylates, curing rates of pure acrylates, stress build-up during cure, comparison with fluorescent probes and rates of thiol and ene consumptions in mixed formulations. In the last study, ene reactions were found to be some 15% faster than thiol reactions, giving rise to a predictive model. Fluorescence probes continue to be widely investigated and developed for monitoring cure rates and microviscosity properties of resins. Blue-shift probes are valuable where diffusion of the non-reacted probe out of the rosslinked regions is of paramount importance for optimum probe response. Here a decrease in probe sensitivity was observed when the size of substituents increased, whereas more reactive substituents increased probe sensitivity. Reactive and non-reactive probes have been developed based on nitro-stilbene.

Gelation processes are also important where a sudden change in rotational processes occurs. with pyrene being an excellent probe. Fluorescent probes in composities can be enhanced through complexation with divalent or trivalent cations close to the particle clusters, while bismaleimides with electron-donating amine moieties exhibit strong self-quenching. The binding of fluorescent probes to glass-fibre composities has been found useful, while, for monofunctional acrylates the cure rate is a single-stage reaction, difunctional systems exhibit a two-stage process. The latter is no doubt due to a second gelation stage. Solvato-chromic and rigidchromic fluorescent probes have been developed based on diarylene and butadiene derivatives of 4-dimethylamino-4'-nitrostilbene.

Spectral analysis of these probes indicates the existence of an excited state more relaxed than the Frank-Condon state owing to a twisted intramolecular charge transfer, favoured by the co-operative effects of donor-acceptor groups. In other work, probe studies have shown that the primary amino group is several times more reactive than a secondary amine group towards an epoxy system. Both FTIR and fluorescence analysis were found to be consistent in gel-glassy rate determinations. Factors which influence cure rates have been numerous.

N-Vinylamides have been found to be more reactive than acyclic amides in the curing of diacrylates, while other studies have investigated the impact of chain transfer on kinetic chain lengths, light intensity and film thickness, shrinkage effects, pressure effects, glycol fragments, stress factors and temperature effects. For furfuryl methacrylate, a new stochastic model has been devised called 'CORUB'.

A statistical analysis of the model's adequacy shows a good correlation between the experimental data and those modelled stochastically. Photocrosslinking of polymers continues to attract interest for property enhancements. For polyethylene the use of anthraquinones gives a higher rate of chain scission than that for crosslinking, while the photocrosslinking of polyethylene fibres is higher in nitrogen, and new alkyl-substituted benzophenones have been found as effective photocrosslinking agents.

Natural rubber has been photocrosslinked through the use of co-reactive methesacrylate groups, while styrene-butadiene systems can be easily photocrosslinked with a variety of initiators. Here low concentrations of photoinitiators can be used for crosslinking. Poly(vinyl alcohol) has been crosslinked through modification with thymine and uracil derivatives. Photo-dimerization was important at wavelengths less than 300 nm, with greater crosslinking the longer the spacing units. Poly(vinyl abietate) also undergoes photodimerization through the abietate groups, while PVK has been crosslinked with a Nd:YAG laser and poly(3-hydroxyalkanoates) can be useful for photolithography micropatterning.

Rapid curing of thiol-ene systems occurs in air and, for thiol-acrylates and methacrylates, the latter components cure faster than hydrogen-atom abstraction from the thiol. Other applications and materials include the monitoring of butyl acrylate diffusion in polystyrene blocks, development of photocurable membranes responsive to ionic surfactants, preparation and photoisomerism of norbornadiene units on polyesters, photocurable sealants for LCD panels, use of fluorescent dyes for the internal microdiagnosis of polymerized 3D microstructures, development of epoxy and methacrylate monomers with chalcone moieties, hybrid sol-gel glasses with Ba titanate and titanium alkoxides for new opto-electronic devices, fabrication of 3D photonic crystals with lasers, light curing of acrylic composities, encapsulants for micro-electronics, monitoring and optimizing UV curable powder coatings, laquers for DVD's, distortions during prototyping, water-based poly(methacrylate)-modified poly(urethanes), photocurable poly(butadienes) with acid and base stability non-wettable fluorinated acrylics, curing of acrylics under water, yellowing of coatings, curing of multifunctional anhydride resins phase structure of curable blends, crosslinkable poly (caprolactones) for biomaterial applications, use of allyl glycidyl ether to prevent oxygen inhibition in curing, 3D mass transfer in curing of multi-component systems, Langmuir-Blodgett films of L-glutamate and crosslinkable polymers with Rosin groups. Poly(urethanes) with siloxane groups exhibit a high degree of segregation of the flexible units from the hard domains compared with acrylicanalogues.

Polymers of acetophenone *O*-acryloyloxime undergo photo-initiated thermal crosslinking when sensitized by benzophenone; post-exposure bakes enhanced hardening. The molecular characteristics of siloxane polymers have been shown to influence cure rates, while polymers with phenacyl groups are photocleaved to give carboxyl groups. The latter may then be further reacted with epoxy functionalities. Water-developable acrylic resins have been made by co-reaction with quaternary ammonium salts of phenolic resins and stable optical gratings have been made fromamorphous azo-benzene containing polymers.

Poly(allyamines) give swellable macroporous gels sensitive to pH, while other developments include the formation of a dianepoxyde and poly(siloxane-acrylics) for litho-applications. Stresses generated during the photocuring of litho systems have been evaluated, while fillers can reduce shrinkage.

Dimethacrylates have been investigated in detail. Evidence has been presented in support of the existence of chain length dependent termination in crosslinking free radical polymerizations. Shifting the kinetic chain lengths towards shorter chains had little effect on multiacrylate systems, but for corresponding methacrylates the kinetics changed significantly. Shifting the kinetic chain lengths towards shorter chains led to the delayed onset of a reaction-diffusioncontrolled termination behaviour, which could influence network formation in systems above their *Tg*.

In another study, photopolymerization kinetics for dimethacrylates showed the presence of two regions, namely, quick and short, and slow and long. Rigid and flexible dimethacrylates have been mixed and studied, and it was found that the latter exhibited faster rates, while in another study the reverse was found, where Bisphenol A resins cured faster than those based on ethylene glycol. Other work has showed that conversion rates are influenced by the initiator concentration and absorption spectrum, and these in turn are influenced by the light intensity.

Coupled with this is the fact that the final properties of the system will depend on the degree of conversion, temeprature and the number of methylene groups in the monomer. Dimethacrylate phosphonates have been made but found to be much less reactive than conventional dimethacrylate systems, whereas for dioxolane derivatives of bisphenol- A no complete ring opening has been found in the presence of a free radical initiator. Hyperbranched dimethacrylate polyesters have been evaluated for use in dental systems, and dendtric polyester acrylates have been made that require no initiator.

Polyimides and polyamides have also attracted some attention. Photosensitive polyimides with benzophenone moieties and hypersensitizing properties have been synthesized and their properties easily controlled for cure. Hyperbranched poly (ether-imides) are useful positive working resists, whereas fluorine containing polyimides have excellent near-UV transparencyand excellent thermal stability. Polyimides with a high content of carboxylic acid groups have enhanced photosensitivity as do resins with dianhydrides. The incorporation of silica enhanced the physical properties of the cured polyimide materials. Hyperbranched aliphatic acrylated poly(amide esters) cure effectively to give thermally stable systems.

Photografting

Poly(acrolein) particles have been photografted with tetramethylcyclotetrasiloxane via cleavage of the Si-O bonds. Thin layers of *N,N* diethylamino) dithiocarbamoylpropyl) (trimethoxy) silane and *N*-isopropylacrylamide have been photografted onto silicone wafers, which exhibit a fast temperature response. Surface modification of poly(imides) has been undertaken with acrylic monomers for potential high tech applications.

Maleic anhydride can be grafted to polyethylene films with high adhesion strength at 80°C, while various acidic gases can grafted onto diffent polymer surfaces to give nitrile, amino and sulfonic acid groups. R are earth metal complexes have been homogenized on polymer surfaces by effectively photografting diketone ligands. Micropatterning on surfaces can be affected provided a photoiniferter is present with acrylic acid giving a better graft than dimethylamino methacrylates. A pore-filling process has been used to produce high-performance composite membranes for selective pervaporation separation of aromatic and non-aromatic solvent vapours. These were based on methacrylate monomers grafted onto poly(acrylonitrile).

Vinyl acetate has been photografted onto polyethylene using benzophenone as a photoinitiator. Here preadsorbing the benzophenone to the polymer gave better graft yields. Vinyl monomers have also been photografted onto silica surfaces and poly(propylene). The surface adhesion of polyethylene has also been enhanced by grafting with methacrylamide as well as enhancing its wettability.

A method of grafting conjugated polymers onto solid substrates by self-assembled layers has been developed. Here poly(fluorene) was deposited onto electrodeposited carbazole. Thiocarbamate groups have also been photografted to copolymer surfaces, as have thiol-terminated polymers and hydrogels onto gold. Other studies of interest include the production of semi-fluorinated alkane thiol self-assembled monolayers with diacetylene units, photofunctional core shell micro-spheres, acrylated polyurethanes grafted onto acrylic copolymers and methyl acrylate grafted poly(acrylonitrile) for separation of benzene-cyclohexane mixtures.

Luminescence and Optical Properties

A number of review articles have appeared covering various aspects of polymer luminescence, optical and conductive properties. These include smart materials based on functionalized polymers, light emitting polymers for visual applications, photophysical characteristics of water soluble polymers, micellar media, diacetylene single crystals, conjugated metal complex polymers, semiconductive polymers, charge-carrier generation, and intra-chain fluorescence quenching studies.

The chemiluminescence (CL) property of polymers continues to attract a steady interest in the field. Much of the work is directed towards investigating and probing the degradation and oxidation processes in polymer materials. The technique of second time-derivative analysis has been developed for studying chemiluminescence profiles of polypropylene-polybutadiene blends. It is concluded that this technique, when used in conjunction with the integrated CL profile, will enable evaluations of the oxidative induction time (OIT) to be made. For the polymer blends, each phase was found to oxidize independently with separate activation energies. The propagation of CL as fronts through stabilized polymer films could be used as a means of measuring the effectiveness of different stabilizers.

The technique was found to be consistent with that undertaken by FTIR analysis. Arrhenius plots have also been obtained for stabilized poly(propylene) films to establish the consistency of measurements and suitable testing conditions. Recycled poly(propylene) films have been shown to exhibit a different CL behaviour to that of conventional virgin material. The former shows threadlike defects as well as a significant change in oxidative stability, which was apparently detectable much earlier than by FTIR and DSC methods. In another study on poly(propylene), the CL intensity has been found to be proportional to the growth in carbonyl species. It was proposed that the CL was due to either energy transfer from the carbonyl species to a more effective emitter or the reaction of peroxides with an oxidation product. Clearly, the nature of the CL species is still elusive.

Kinetic evaluations on pre-oxidized poly(propylene) has, however, shown that the rate constants for CL growth match those for fast decomposing hydroperoxides. Similar results have been obtained on the CL analysis of low density polyethylene, where it has been used to analyse trace levels of residual hydroperoxides. Characterization of oxidation rates of polymers and analysis of initiation sites and stress levels have been covered in a number of studies. Studies on poly(2,6-dimethyl-1,4-phenylene oxide) and polyesters have shown that the CL behaviour is pre-determined by the oxidation levels and presence of oxygen during measurement. Hydroperoxide sites are found to be responsible for the thermally induced CL in polyesters, with the nature of the aliphatic backbone playing a major role in controlling the CL behaviour.

This is further borne out by a study on cellulose, which showed that the CL was dependent not only on the presence of oxygen but also on moisture, no doubt owing to the formation of hydrogen peroxide. In the advanced stages of degradation the CL showed a dependence upon alkali-earth metal ions. Here the metal cations could potentially complex with the glycosidic groups. Concentrations of hydroperoxides have also been evaluated in polyester-melamine resins by CL while CCD devices have been found useful for the detection of CL from PVC formulations. In other applications CL can also detect polyphenols in fabrics, and an electrochemical technique has been developed for the detection of coating integrity based on CL emission.

Violet CL emission has been observed from a novel condensation poly(imide) and from fluorophores containing distyrylarylene unit. CL has also been observed from light irradiated wood surfaces and found to be correlated with discolouration. The electrical breakdown of epoxy resins gives rise to electroluminescence, while peroxides are associated with their thermally induced oxidative generation of CL. The radiothermoluminescence from different polymers is also dependent upon oxygen pressure.

Polyacetylenes continue to attract some interest. Using Raman scattering, the primary photoexcitation in trans-polyacetylene and polydiacetylene is considered to be a breather exciton, a bound state of an exciton and a localized non-linear lattice excitation associated with stretching vibrations of the polymer backbone. Phenyl substituted polyacetylenes have been evaluated using multi reference singles and doubles CI calculations. Apparently, strong luminescence is due to the fact that the 2Ag two-photon state lies above the optical 1Bu state, in spite of the polyene backbone.

Reversed energy ordering in these polymers, as compared to other linear types, is a signature of redu-ced effective electron correlations. The optical gap of the substituted materials is smaller than that of polyenes of the same length. Also of interest is the observation that the dark conductivity of poly(diacetylene) is enhanced after laser irradiation and that this effect may be due to the growth of surface oxidation products. On the other hand, electrically induced molecular re-orientation can fine-tune the photoluminescence from poly(acetylenes), while benzyl-carbazole substituted poly(acetylenes) exhibit much weaker emission. Anthryl substituted poly(acetylene) has been made and is black in colour with a visible cut-off at 780 nm, while the photo-luminescence of 1-naphthyloxy-1-alkynes gives stronger emission in the monomeric than the polymeric state. The polymerization kinetics of ethylmorpholine pentacosa-diynoic amide were measured in floating and supported monolayers at an air-water interface. On irradiation there was an observed blue-to-red polymer colour change with quite distinct levels of fluorescence due to a rearrangement in the packing of the side-chains along the polymer backbone.

Studies on polymer blends have provided valuable information on miscibility. Multiphase behaviour has been observed in copolymer blends of dimethacrylamide and methacrylic acid. Although three main phases were observed, transitions among different phases were discrete with hysteresis loops. Fluorescence analysis indicated the coexistence of hydrophilic and less hydrophilic domains, with their fraction depending upon pH. The monitoring of in-situ blends of methacrylate polymers was found to be inconsistent under shear induced coalescence, whereas, in the absence of shear energy, transfer measurements with donor-acceptor probes were in good agreement. Blend miscibilities of poly(fluorenes) have also been evaluated, as have blends of PPV's with poly(ethylene oxide).

Excimer formation has been used to determine the miscibility of poly(vinylcarbazole) with PMMA, while complexes of poly(acrylic acid) and poly (vinylpyrrolidone) have been found to be quite stable owing to associated hydrogen bonding. Poly(vinyl alcohol) and poly(vinyl acetate) blends have been prepared on a poly(ethylene) surface and found to give different domains, depending upon the degree of polymer interpenetration.

Other studies have looked at the morphology of epoxy-poly(organosiloxane) mixtures where there was some diffusion restrictions and hydrogen bonded poly(acrylics) using the photocontrolled pitch of a cholesteric helix. In the latter, blends with a chiral nematic phase were formed over a wide temperature range and for films with planar orientation a selective light reflection in the visible and IR spectral regions is observed, with a temperature and UV irradiation time dependence.

The properties of dendritic structures of various types have attracted interest from their photophysical and photochemical aspects. Doughnut-like structures have been observed from methyl ethers and phenylenediacetylene derivatives up to 150 nm in height and 8 microns in diameter. Here no correlations were found between the degree of orientation and the formation of doughnut structures.

Fluorescent poly(benzyl ether) dendrimers with isobutenyl and naphthoate groups have been made with effective light-harvesting properties, as have similar polymers but with a rhodium (III) porphyrin focal core. Phenylacetylene copolymers with emission quantum yields of up to 0.97 have been made, while dendritic molecules with a 1,1'-binaphthyl core have been found useful for enantioselective recognition of chiral organic molecules. A series of star shaped biocompatible polyesters and polylactides with ruthenium(II) tri-bipyridyl cores have been found to exhibit reversible thermochromic bleaching as well as reactions to external chemical stimuli for sensing applications. Dendritic polyesters have been made with high methacrylate functionality and found to exhibit superior properties after photocuring than linear structures.

Cure speeds, however, are significantly slower. For a series of firstgeneration polyphenylene dendrimers substituted with spatially well-separated peryleneimide chromophores and a terryleneimide trapping chromophore at the rim, over 95% of the total energy absorbed by the molecule is trapped by the outer structures by a FRET process. Ethylenediamine with methyl acrylate dendrimers and an anthracene moiety in the core have been developed that undergo a reversible photocyclization process.

The photoalignment of 'allazobenzene' dendrimers has been investigated, whereas polyamidoamine dendrimers with azobenzene groups undergo reversible *trans-cis* isomerism. Dendrimers with a periphery of dibutylamino groups are good donors, while dendrimers with spironaphthoxazine groups undergo not only photoresponsive changes but also reversible complexation with metal ions. Eosin-based dendrimers give less singlet oxygen than the unbound molecule, while pressure affects the luminescence from methoxyphenacetylene dendrimers. Water soluble dendrimers of poly(sulfon-amines) have been found to be highly sensitive to Fe(III) ions with little interference, whereas the caging of drug molecules can be undertaken via the use of dendrimers with peripheral carboxy-2-nitrobenzyl groups. New methods have been developed for polyamidoamine, stilbene and polyphenylene dendrimers. Aggregation effects have been found to increase from a series of diblock to star-block methacrylate copolymers, and poly(triazine esters) are photolabile.

Hyperbranched poly(ether ketones) with 1,3,5-triazine moieties are amphiphilic, forming self-aggregates at a specific concentration.

Similar materials have been made from poly(ethylene oxide) with hydrophobic dendritic carbosilane groups.

Block copolymers of third generation could not be dispersed in water, whereas the first and second generations formed micelles with a multi-core lower viscosity centre. Polyamidoamine starburst dendrimers have been shown to form a tight non-polar association, which becomes more densly packed with increasing generation. Only weak fluorescence is observed, possibly due to the np* character of the lowest excited singlet state of molecules.

Porphyrin and pyrene molecules have been encapsulated within polymeric shells by ring-opening e-caprolactone. With increasing chain length the core is isolated further, thus providing a useful methodology for monitoring the effectiveness of encapsulation techniques. Other studies of interest include the development of star shaped polypeptides, polyelectrolytes based on poly(*t*-butyl methacrylate) precursors, fullerenes with iniferter *N,N*diethyldithiocarbamate groups, fullerene star shaped polystyrenes and porphyrin containing polyamides.

Finally, by careful design of dendritic molecules the physical properties of the self-assembled structure can be tuned in such a way as to phototrigger micromanipulation, giving rise to the formation of hard spheres. This possibility leads to the bottom-up synthesis of micrometer sized objects through a combination of covalent synthesis and supra-molecular organization.

Photochromic systems continue to be highly active. Photoinduced rearrangements readily occur from the 'trans' quinone forms to the 'ana' quinone forms for a styrene derived phenoxyanthraquinone in both solution and film. Polyesters with norbornadiene groups photoisomerize smoothly to the quadricyclane form, as does a norbornene derived polymer with phenoxyquinone groups.

Spyropyrans in polymer matrices have significance. Pyran doped polymers have particular value in holographic recording media, with a number of systems having been developed based on spirocyclopropenes, hydroxy-9*H*-xanthen-9-ones, bisindolinobenzospiropyrans linked to thio and carbonyl groups, dicyanomethylenefulgides, new spiropyrans with thermo and photochromic properties and photochromic acrylates with biphenylene and spiro-oxazine moieties with chiral substituents. In the latter case, the homopolymers did not exhibit mesophases owing to steric hindrance between the side groups.

Azostructures continue to be one of the most studied, with a number simple dopant studies and the grafting of azo dye chromophores to effect optical isomerism. Many other studies deal with co-reacted azo-type funtionalities. Hydrophobically modified poly(acrylates) with azobenzene groups have been observed to give unusually high viscosities at certain critical concentrations owing to reversible crosslinks. On irradiation the *trans* to *cis* isomerization gave rise to viscosity changes due to inter-polymer associations.

Azo group labelled polyesters also undergo rapid reversible *trans-cis* isomerization, as do poly(malonic esters) with *p*-cyanobenzene groups. In this case unexpected photoinduced birefringence was observed at two laser pumpings of and 365 nm, which were attributed to some other unassigned process and not solely isomerization. PMMA with azo groups also exhibit birefringence with a dependence upon the polarization angle of the laser pump beam and probe beam.

Here a model was developed to describe the geometry of the isomerism and to correlate its treatment with the writing-erasing processes of the photoinduce birefringence. Activation energies and hole-buring proceses have also been measured for poly(acrylics) with azo chromophores. Photo and thermally reversible isomerization has also been evaluated in amphiphilic copolymers of 4-(4-dodecylphenylazo) benzoic acid, while in another study relief structures have been imaged by coupling azobenzene groups to urethane-based polymer systems. For blends of LC polymers of alkylhydroxybenzoic acid with 4(4- pyridyloyl) oxyazobenzene, an increase in the concentration of photochromic groups favoured a change in the phase state of the material.

In principle it was shown that hydrogen-bonded photochromic blends may be used as optically active media for data recording. For copolymers of*N*-(2-hydroxypropyl) methacrylamide with 6-[4-(4-methoxy-phenylazo) phenyl-hexyl methacrylate, the azo functionalities were found to self-organize into dimeric aggregates, while aggregate formation in the PMMA-azobenzene system is highly dependent upon thermal treatment. In this case, the first writing cycle of the aggregated polymer film showed a steep increase in photoinduced birefringence followed by a gradual drop to a certain asymptotic value over longer irradiation periods with polarized light. New write-erase chiral photochromic cholesteric shaped acrylic copolymers have been made with azobenzene groups and chiral menthol or menthone functionalities.

The cholesteric helix untwists on irradiation with a corresponding shift in light reflection to longer wavelengths. This process is thermally reversible with the menthyl system. Ionomeric polyurethanes have been made that can undergo an ion exchange with 3(*p*-oxyazobenzene)-sodiumpropane- sulfonate groups while a theoretical model has been developed that describes a temperature field which is established within a film of isotropic azobenzene LC side-chain polymer under linear polarized laser beam irradiation.

The minimum light intensity required to induce a photo-induced phase transition as well as an offset photo-alignment temperature were calculated. Isothermal luminescence has been observed from PMMA films doped with aromatic amines which had been ionized. On cooling the amine radical cations recombine with trapped electrons emitting light. Similar processes were developed with compounds based on 3-methyl-1-benzothiophen-2-yl moieties, while those based on and *N*-allyl-maleimide did not work. Helical structures have been developed based on poly(fluorenes), while the magnetic properties of photochromic stilbenes with diarylethene units have been investigated.

Here the magnetic properties of the molecules are interesting since they can be switched from ferromagnetic to non-magnetic with some ring opening in the isomers also taking place. Copolymers of 3-tetradecylthiophene and 3-*p*-trifluoromethylphenylthiophene exhibit reversible thermal behaviour. Thermal motions of the side chains result in extended conjugation to the phenyl groups along the polymer backbone. Above 70°C the conjugation begins to break down, with the fluoro derivative exhibiting blue shifted emission due to the electron withdrawing effect.

Thermally induced phase transitions have been studied for 5,7-dodecadiyne-1,12-diol bis[phenylcarbamate] as well as diarylbutadienes. Hinged poly(peptides) have also been made which show uniform *cis* and *trans* isomerism for photochromic moieties that are close to the terminus of the polymer chain. An analytical theory has also been developed for couples undergoing photoi-somerization and photoorientation, both for individualized and spectrally overlapping isomers. When diarylethene and spiropyrans are oriented by polarized light, the apparent optical orientation changes sign for the UV versus the visible photochemical transitions. Other studies on chromic materials include polymers with azobenzene side chains, spiropyran doped polystyrene, triazine-stilbene fluorescent brighteners, pyrocarbazoles and LC polymer microspheres. LC polymers follow on with a similar active interest. A number of LC polyesters have been prepared with 1,3,4-thiadiazole and 4-hydroxybenzoic acid and 6-hydroxy-2-naphthoic acid.

These polymers form stable enantiotropic smectic and nematic phases. Photoinduced alignment studies have been undertaken on several polyimides, with fluorine and chalcone groups 4-styrylpyridine groups and methyl red dye as dopants. Here pre-tilt angles were found to be related to the degree of imidization, so by adjusting the chemistry of polyimide formation the pre-tilt angle can be varied continuously. Varying the structure of the polyimide produces LC's which will align either parallel or perpendicular to the polarization direction of the UV light. Significant energy transfer (FRET) has been observed between a LC fluorene-based polymer and a non- LC polymer. In this case a low threshold photo-pumped amplified spontaneous emission was observed in a waveguide device fabricated with the blend. This low threshold emission originates from a largely reduced self-absorption loss by the efficient energy transfer in the photoluminescent polymer blend with LC properties. PMMA with terphenyl side groups has been made, as have polyethers with stilbene side groups.

Both types exhibit LC properties, the latter having bitropic properties, with low and wide phasetransition temperatures as well as good thermal stability. New chloesteric polymer materials possessing double photochromism have been made from copolymers with phenylbenzoate hematogenic and spyropyran side groups.

Irradiation and annealing of planarly-oriented films made of these systems leads to the untwisting of a cholesteric helix, inducing an irreversible shift of selctive light refletion to the long-wave region owing to the *E-Z* isomerism. The thermally reversible spiropyran gives a separate process but one which could be used in conjunction with the *E-Z* process for photoelectronic recording systems. The LC properties of AB block copolymers of polystyrene and poly(isoprene) with 4-cyanophenylazobenzene side groups is influenced by the microstructure of the materials related to the phase separated coils. Azo grafted ladder-like polysilasesquioxanes normally used for command surface fabrication have an improved photodriven response.

The response steepness of an LC cell was found to be two orders of magnitude greater than previous work with an optimum response at 10 mole-% of azo content. The LC alignment in films of polyimide with cinnamoyl and coumarin groups has been found to be induced by the photodimerization processes rather than any *trans-cis* isomerization while the photoinduced alignment of LC polymers with azobenzene groups has been optimized. It was found that the response could be enhanced by chemical tailoring of the polymer structure and physically modifying the alignment procedure.

New comb-shaped photochromic and LC ionomers have been made by protonating 4-(10-diethylaminodecyloxy)azobenzene through the carboxylic acid groups of functionalized polymers. Here the growth and decay of induced birefringence for ionomers and photochromic polymers were found to obey similar laws. A series of polyacetylenes containing 4-(*trans-n*-alkylcyclohexanylcarbonyloxy) phenyl-4-alkynyloxybenzoate side groups have been found to exhibit nematic, smectic A and smectic B phases, while the monomers showed only a nematic phase. The polymers gave green-blue luminescence and had an interdigitated bilayer structure.

Homogeneous alignment has been maintained in a series of three ring phenylacetylenes, while a series of acetyl (ethyl) cellulose-acrylic systems reflects one hand of circularly polarized light in a narrow region of wavelengths. LC transitions have been probed via the use of fluorsecent molecules, and nematic diepoxides have been developed for microstructured devices. Other LC devices include a series of multifunctional chiral photochromic materials and antiferroelectric diacrylate gels.

UV irradiation has been used to control and tune the properties of emulsion polymers with azobenzene chromophores, and phase separation has been examined and calculated for LC unsaturated oligoester resins. The area of LED polymers is the most actively studied in the field.

A number of articles have dealt with various aspects of LED's such as their preparation, improvements, ladder structures and engineering aspects. The excited state characteristics of these polymers continue to attract much theoretical and experimental consideration. From configuration-interaction calculations on PPV's, intersystem crossinking from the singlet to the triplet manifold is associated with spin-orbit coupling, giving rise to both singlet and triplet wave functions. Using femtosecond spectroscopy the optical emission form PPV's is described by exponential functions, while transient absoption measurements show that charge-carrier generation occurs 100 fs after excitation.

The carrier generation is also influenced strongly by the strength of interchain interactions. Structural changes in a single molecule over periods of seconds have been measured by monitoring fluctuations in the wavelength of emission spectra using total internal fluorescence microscopy, while quadratic coupling between ground and excited states has also been measured by fluorescence analysis. The latter results from twisting about the oligomer axis, which is free in the ground state but strongly constrained in the excited state as a planar configuration.

A theoretical description of excitions in PPV's has been obtained using the electron-hole Bethe-Salpeter equation, and the Peierls model used to solve the energy and soliton structure. With the Green function method the extent of polarization can be determined, while other work has shown that quenching processes are significant in conjugated polymers, with solid-state materials exhibiting over 50% more intense emission than corresponding solutions. In contrast, solutions give stronger emissions owing to interchain quenching process in the solid phase. Energy migration in PPV's is not fully accounted for by the FRET mechanism, with molecular aggregation having a profound effect on the excited state properties of the polymers.

The way in which polymer films are cast is also very important. This can influence the nature of any inter-chain interactions and subsequent electroluminescence efficiency. Of particular relevance is the study of inter-chain aggregates in polythiophenes; time-resolved analysis has identified the emissions from intra and inter-chain aggregates.

Dynamic quenching of luminescence in films with dense packing is mainly due to excitation energy transfer to aggregates. Here the aggregate contribution to the total luminescence in polythiophene films with dense chain packing is about 50% and is virtually independent of temperature.

For PPV's evidence is also presented that photoexcitation of aggregated polymer chains results in the creation of interchain non-emissive polaron pairs in large quantum yield. This effect accounts for the observed lower quantum yields in the solid state and supports the picture of polaron pairs recombining to form the ground state and not the corresponding singlet and triplet states. On the same theme, chain conformation strongly influences the nature of the fluorescence from PPV's. Thus, while extended chains give emission from multiple segments, folded chains give emission from only a few sites. The tightly folded chains enable the polymer to funnel excitons to highly aggregated low energy regions via a 3D FRET. This strong intra-chain coupling causes the polymer to mimic the photophysical behaviour of single chromophores, leading to localized emissions nad fluorescence intermittency. Polymer molecules with specifically designed steric hindrance for backbone contact show higher resistance to solvent induced interactions between segments. Non-solvents force the polymer chains apart causing stacking and aggregation. These buckled chains show red-shifted excimer fluorescence. Such effects can be detrimental to device performance.

Inter-chain coupling of PPV derivatives in solution is more predominant for *cis* than for *trans* isomers, with nanotubes offering the best performance for electroluminescent devices. New phenothiazine-quinoline copolymers have smaller band gaps than their corresponding homopolymers, giving blue and red emission bands associated with intramolecular excitons and intermolecular excimers. On a similar theme, interfaces of polyquinolines and polythiophenes exhibit strong quenching due to electron transfer from the quinoline to the thiophene groups. A new weak red emission was observed associated with an unreported exciplex site at the interface. It was suggested that such interaces could have valuable application in photovoltaic devices. Electroactive poly(vinyl diphenyl-quinoline) has been made and found to be phototunable in acidic media owing to intramolecular excimers. Apparently, in neutral or thin film media, no excimer formation was observed giving rise to electroluminescent sensors.

Single layers of carbazole and fluorene containing PPV's have been found to give rise to pure blue emission, and stronger electroluminescence has been observed from D-A bipolar oligomers of PVK with dithienothiophene. Similar results were obtained with alternating thienylene-phenylene copolymers giving red-shifted emission spectra. Varying the thiophenoxy precursor chain length in the PPV is another means of altering the band gap. Another method of enhancing the charge-carrier transport has been to develop a two-step synthesis for introducing electron accepting sulfone groups, while more effective polymers have been developed using polymers with dithiafulvene and bithiazole groups. Thiophene-*p*-phenylene co-oligomers have been molecularly modified to enhance emission tuning and control their epitaxial crystallization during vapour deposition of thin films. Here films with highly polarized emission were formed, providing useful materials for isotropic applications. Magnetically designed polymers have also been developed to produce dichroic luminescence and electrical anisotropy. These were useful for LC devices. Bipyridyl molecules have been incorporated into water soluble PPV's to examine the emission quenching processes. Quenching increases with increasing charge.

Substitution of solublizing groups, especially those based on the alkoxy functionality, continues to be researched. To avoid the usual problem of phase separation between conjugated and ionic polymer in LED electrochemical cells, flexible poly(ethylene oxide) blocks have been successfully introduced into PPV.

Emission properties were retained while for a series of alkyl-substituted PPV's the emission quantum yield decreases with increasing chain length. For increasing alkyl chain size there is also a corresponding blue shift in the emission maximum and decrease in electronic delocalization along the chain. On the other hand, enhanced green electroluminescence has been obtained through the development of blends of PVK with alkoxy substituted m-phenylene PPV's. For octyloxy derivatives exciton bands were found to decrease in intensity, while PPV's with 2,3-substituted dibutoxy or dimethyloctylsilyl side-chains gave double layer emitting devices with eletroluminescence efficiency and low turn-on voltages.

Blends of methoxy and dioctyloxy PPV's have been found to exhibit amplified spontaneous emission, while copolymers with oligo(ethylene oxide) side chains have been designed to combine high luminescence efficiency with ion-co-ordination ability. PPV's have been found to retain a rod-like structure in poly(styrene), and novel LED electrochemical cells have been made by copolymerization of PPV's with crownethers. An LC PPV with two octyloxy-pentylcyclohexyl-phenyloxy side chains hase been made and found to exhibit effective electroluminescence-device characteristics in a sandwich structure test.

Fluorene structures have also attracted some interest. Hyperbranched PPV's with fluorene moieties exhibit short wavelength blue emission with high quantum yield and good thermal stability. Similar copolymers have been found to give much stronger emissions in solution compared with the solid state owing to the latter forming strong aggregates.

A colourfast poly-2,8-(indenofluoreneco-anthracene) copolymer has been made which unfortunately undergoes ageing on storage, causing a reduction in emission intensity, with similar problems occurring in other poly(fluorenes) where a small number of dimer sites have been identified. The latter case causes exciton quenching and excimer formation. Other trap free systems based on poly(2,7-(9-fluorenone) have been developed, while in another study dialkylfluorene units give polymers with enhanced electron affinity and ionization potential.

To combat the problem of aggregation, copolymers have been developed based on polyfluorene and carbazole units. These disordered polymers gave strong emission with good thermal stability. Polyfluorenes coupled with carbazole units have also been combined with acetylene units and PPV's with phenylphosphne oxide moieties. The latter have well defined emission spectra in the blue and red regions of the spectrum. Highly effective *trans*-poly (9,9-di-*n*-octylfluorene-2,7-vinylene) has also been made by acyclic diene metathesis polymerization using Mo catalysts.

PPV's with oxadiazole groups have also been developed for enhanced LED applications. A biphenyl-oxadiazole derivative has been found to exhibit effective electron-hole transport with good photostability, while a bipyridyl derivative was found to be unstable. Oxadiazole polymers have been found to be excellent energy donors in blends with PPV-carbazole derivatives, giving strong blue-shifted emission. Similar polymers with carbazole and oxadiazole units have been made with green emissions at 530 and 540 nm, respectively. The former had the best device performance, with a low turn-on power.

Other oxadiazole polymers with differing content of 2-ethylhexyloxy pendant groups have been found to be stronger emitters than PPV itself, as well as being effective EL devices when layered with In-Sn -oxide coated glass. Other polymers with green emitting oxadiazole and 2,4,5-triphenylimidazole chromophores have also been synthesized. The photophysical characteristics of various structural types of PPV's have been investigated and collated. These include some cyano derivatives and block copolymers. Although near unity quantum efficiencies were observed, excimer formation is quite evident in thin coated films. PPV's with dioctyloxy groups behave like a collection of semi-independent excitons with dipoles arranged in the same direction.

DC background measurements, however, indicated a few misalignments. Tunable PPV's have been prepared for optoelectronic devices, while a reduction in temperature enhances the EL intensity. Vacuum deposition and doping with iron have also been found to be useful ways of enhancing the luminescence from PPV's. High stereoregularity in poly (phenylacetylenes) gives polymers with specific emissions that are independent of solution concentration when compared to irregular polymers.

The integrated fluorescence lifetime of polycrystalline PPV's decreases with increasing temperature, with a corresponding increase in anisotropic decay. In 2- methoxy-5-hexyloxy-substituted PPV's excimer emission is evident and there is no temperature dependence on the fluorescence. Localized chain interactions are responsible for these differences. Similar results were obtained on the same alkoxy derivatives of PPV after spin coating solid films, where aggregate trapping was more important. Other studies include ionic quenching and photoexcitation dynamics in ladder type structures. Other PPV's have been made with stilbene-*N*-alkylpyridinium chromophores, where the nature of the alkyl group strongly influences the emission quantum yield, with the most efficient being those with long aliphatics. *Z-E* isomerism is also important in PPV's, with azo groups enhancing optical anisotropy owing the side-chain reorientation. PPV's with side-chain vinyl groups have also been prepared and characterized, showing fibrillar and network type morphology.

Delayed fluorescence from 1.5% of triplets due to T-T annihilation has been measured for 2-methoxy PPV, while enhanced emission was obtained after copolymerizing PPV with the 1-methoxy derivative. Organometallic modules strongly influence the emission properties of PPV's, while in *m*-phenylene stacking of PPV's is prevented. Alternating rigid block copolymers of PPV's with poly(ethylene glycol) units have been made to enhance solubilization, whereas a new method for PPV synthesis has been devised via a Diels-Alder reaction. PPV's with amine groups have been investigated. Alternating copolymers of the methoxy-PPV, vinylcarbazole and a triphenylamine-PPV give very high luminescence intensities. Hyperbranched PPV's with triphenylamine holetransporting groups give good films with high luminescence quantum yields in the red region.

PPV's with alternating triphenylamine groups give strong green luminescence but high turn-on voltage, while the polymerization of PPV's at low temperatures does not alter their photoluminescence efficiency. The incorporation of alternate *meta*- and *para*-phenylenes in the PPV chain gives a well blue-shifted emission owing to reduced conjugation lengths. Intraand intermolecular aggregates have been quantified in PPV's, with the former being associated with the luminescence. Monfunctional oligo-PV's with *s*triazine units aggregate in chiral stacks.

Ionic PPV's and derivatives have been examined. This includes the lyotropic properties of LC poly(pyridinium) salts in protic and aprotic solvents, where films were formed easily and luminescence arose in methanol and DMSO. For model systems the presence of potassium nitrate influences the state of aggregation in micellar media, while oligomers of PPV's have been found to behave as more effective sensors to analyte quenchers than their polymeranalogues. Carboxy-substituted PPV's have been made with water solubility, giving narrow photoluminescence spectra.

Such structures were effective for multilayer self-assembly techniques. In such systems excited state dynamics can be effectively controlled even on the nanometre scale. Energy-transfer processes have been examined in detail in such systems with dopands such as Cu phthalocyanine. This allows the development of controlled photovoltaic devices. A novel soluble derivative of PPV has also been made. Modified PPV's are also being investigated for more practical applications. This includes doping for example, in polymer-silica hybrids. Results so far indicate the enhanced formation of J-aggregates, a blue-shift in the emission of some 80 nm, a marked increase in efficiency and higher photostability. Similar work has been undertaken on binding into clay. In this case, a 2-D lamellar composite structure efficiently enhanced the electron-holerecombination rate by blocking charge decay. Additionally, the isolation of the polymer chains within a confined geometry by intercalation prevents the excitons from finding low energy trap sites.

As mentioned above, PPV's with ethylene glycol groups have improved ionic solvation, thus electrodes with Ca cathodes can achieve high luminance intensity for electrochemical cell behaviour. Polyether-based PPV's with high luminance and good solvent solubility have also been developed, while multicolour devices have been made based on PVK.

Fluorescence quenching in diphenyl-substituted PPV's is believed to be due to exciton-exciton annihilation flowed by polaron-pair formation and then further annihiliation. In order to reduce chromophore aggregation in PPV's and enhance emission intensity, planarization has been calculated to give modest shifts of 20-30 nm, which can also preserve the vibronic structure of the monomer.

The introduction of diphenylmaleimide structures introduces kinks into the PPV chains and gives twin emissions at 413 and 530 nm. Pendant fullerenes, on the other hand, give two signals due to photoinduced transfer of electrons between the conjugated chains as donors and dangling fullerenes as acceptors. Fully functionalized fullerene PPV's have also been made that are solvent soluble. In poor solvents, cyano-substituted PPV's give red-shifted emission spectra due to increasing inter-chain states, while *cis* and *trans* forms give completely different spectral emissions. Very stable tetra-fluorine PPV derivatives have been made and found to be good emitters with bands at 485 and 515 nm. In contrast to many fluorescence studies, other workers have ascertained the importance of the triplet state through characterization of the phosphorescence emission form PPV derivatives. Metal chelation of PPV's through reaction with 2,2-bipyridyl groups attenuates electron delocalization.

The introduction of siliconbased alternating units also disrupts the electron conjugation in PPV's. Nevertheless, these materials were more easily processed and gave distinct emission spectra dependent upon the turn-on voltage. Other LED polymers include bifuryl-polyimides with emission yields of up to 15%, 1,3,5-tris-styrylbenzenes, phenyl-capped poly(anilines), poly(anilines), 1,4-bis(3'-methyl-6-octylstyryl) benzene, ionic poly(ethylene oxide), PVK blends, 2,2'-bipyridyl polymers, carbazolyl-methacrylates, Pt-doped ethynylenes, poly(pyridine vinylene), Eu-complexed PVK, a poly(naphthyridine-2,6-diyl) complex with $Ru(bpy)_2^{2+}$, and poly(heteroaromatic oxadiazoles), Polythiophenes have also been widely investigated as LED's. Triplet energy migration has been observed to the longest conjugated segments in 3,4-dibutylthiophenes, while, in poly(3-butylthiophene) layers, mobility has been found to be unaffected by thickness.

The morphology of poly(thiophene) derivatives depends upon the nature of the alkyl substituents and and chain length. Here the shorter the chain length the greater the emission intensity owing to a decrease in hopping distance. These materials, however, were highly sensitive to oxygen quenching. Other studies on poly(substituted thiophenes) using near-field fluorescence microscopy have shown that protruding domains are blue-shifted compared with flat areas, with further blue shifts as the irradiation time progressed in the scanning microscope. For a series of poly(3-alkylthiophenes), long chains are considered to enhance the processability of the polymer as well as its emission quantum efficiency. These polymers gave strong emission at 520 nm in chloroform and 570 nm in the solid film. Poly(2-alkyl-benzimidazole)-altthiophenes, on the other hand, give twin emission peaks at 450 and 525 nm. For poly(3-(2-methoxy-5-octylphenyl) thiophene), highly emissive microcrystalline structures are formed in poor solvents, whereas a series of tetraacetylenic thiophenes give emissions varying from indigo blue through to reddish oranges.

Time resolved analysis of the photoluminescence from polythiophenes has indicated a two component decay process, while a series of oligothiophenes with benzyl ether dendrons form well-defined nano-aggregates, in cooled solutions. The extent of aggregation is reversibly controlled over a small temperature range, with strong bathochromic shifts in the aggregated phase. This suggests that interchain delocalization of the singlet and triplet photoexcitations occurs within the dendritic nanoassemblies.

Highly regioregular poly(3-alkylthiophenes) have been made and found to give strong red-orange emission at high conjugation lengths, whereas new types of poly(thiophenes) have been developed which emit over the entire visible range of wavelengths. Carboxylic acid substituted poly(thiophenes) exhibit much redshifted, higher emission intensities in self-assembled layers than in cast films owing to fewer excimer-forming sites.

Light-emitting properties have also been controlled by copolymerizing with cyclodiboranes and fluorene units. In the latter, increasing the oligomer chain length or adding carbonyl groups at the chain ends red-shifts the spectra owing to increased electronic delocalization along the molecular frame. Making polyester derivatives had little effect on the spectral properties.

Hyperbranched poly(arylenes) with thiophene units give strong blue emission and little evidence for aggregation. Dithienothiophenes have been solubilized in protic solvents while at the same time having good non-polar solvent solubility. They also give rise to strong blue emission. Poly(3-octylthiophenes) give red-shifted emission spectra and a lower operating voltage.

New silicon-based poly(thiophenes) can be easily spin-cast with a red-shifted green emission at 520 nm. Other polythiophenes are tetracyclohexyl-3",4"-dihexyl-2,2':5',2":5",2"':5"',2""-quinquethiophene-1",1"-dioxide derivatives, alkoxy-substituted derivatives, terminal-fullerene derivatives and co-fullerene derivatives and poly(3-thiopheneacetic acid).

Dope polymers have been widely investigated as probes of their environment. Pyromethene dyes in acrylic polymers have been studied, as have quinoline derivatives in PVC and HDPE. The latter was supposedly used as a marker for optical identification, but in fact its spectral properties are virtually independent of the environment. Water uptake in polyurethanes has been monitored through fluorescence probes, by and super phosphorescent additives developed for plastics that are better than Zn-sulfide. Conductive patterns have been developed in layers of poly(octamethylene-9,10-anthrylene) through selective irradiation via masks.

Subsequent exposure of the films to iodine vapour gave films with conductive and non-conductive areas, which remain quite stable for periods of months. The spectral characteristics of Oxazine-720 in PVKdepends upon the nature of the doping process. The PVK emission spectra are red-shifted, and there is an increase in intensity with increasing dopant concentration. The distribution of fluorescence lifetimes of embedded dyes in polymer films is shifted to lower values as the film thickness increases. This effect is associated with simple electromagnetic arguments related to the boundary conditions at the interfaces of the polymer film with air and glass. Other probes of interest include 3-coumarin derivatives, polyaniline doped with $H_{3-n}PMo_{12-n}O_{40}$, cellullose esters doped with rhodamine B and fullerene doped polymers.

In the latter case continuous laser irradiation results in enhanced blue-shifted emission. Determination of matrix properties via probes is also useful and in some depth. The thermoresponsive behaviour of microgels based on poly(acrylamide) has been monitored through a pyrene probe. It was found that the use of dimethylacrylamide as a hydrophobic modifier raises the lower critical solution temperature of the microgel. Time-resolved analysis on the nanoparticles confirmed the rapid reversibility of the smart thermal response via a two-stage process. The viability of photon-event counting imaging has been applied to fluorescence studies on dyes such as Rhodamine B in sol-gels based on tetraethylorthosilicate. Here cracks or other defects could be easily identified in the gel. The swelling of PMMA gels has also been monitored by a fast transient fluorescence technique.

Pyrene lifetimes in the gel were used to measure activation energies for the gel swelling process. The technique has also been extended to studying the helical forms of isotactic polystyrene in *cis*- and *trans*-decalin, as well as polymer temperatures during capilliary rheometry.

Pyrene as a probe has also been used to measure the polarity and stability of poly(urethane) micelles. Silica particle growths during sol-gel polymerization have been measured via Rhodamine 6G. Using fluorescence anisotropy, both microviscosity and hydrodynamics can be monitored. The change in the nature of catalyst particles during olefin polymerization on silica can also be monitored via pyrene probes. The same probe has been used to measure molecular relaxation processes in PMMA-silica matrices. There are extensive hydrogenbonding interactions between the ester carbonyl groups of the PMMA and the silicic acid pore surface.

There is also a reduction in the chain entanglements of the polymer and an increase in the free volume due to pore-directed polymerizations. This effect releases several steric hindrances. TICT states of molecular probes also give useful information on chain conformations in polymers. Both free and bound probes will give rise to twin emission peaks, where the close vicinity of the polymer chain is likely to influence the distribution of conformers. Following on, several studies have appeared using pyrene tagging methods as molecular probes. In pyrene end-labelled poly(dimethylsiloxane) there are at least two types of interactions: constrained and unconstrained. The former is due to peptide interactions and precludes excimer formation, while the second results in classic excimer formation. The extent of pyrene interactions have been measured in supercritical carbon dioxide media, and are found to be closer than when dissolved in good solvents like toluene. In a similar way, the intermolecular associations of an oligoisobutylene with a maleic anhydride function at one of its ends has been investigated in *n*-hexane through the fluorescence analysis of pyrene attached onto the MA moiety via an imide bond. At high oligomer concentrations, more than 95% of the oligomer displaying an amide bond is associated. Without the amide bond, this reduces to 60%. In polar solvents, however, this ratio is further reduced owing to solvent associations. Four generations of poly(aryl ether) monodendrons with pyrene at the focus have been made, where structural changes between the generations of dendrons depend upon the nature of the solvent.

From model studies it was suggested that the increased density of larger monodendrons provides a more efficient barrier to the diffusion of oxygen. Dynamic and static photophysical measurements on pyrene attached to a polyethylene: poly(butylmethacrylate)-co-poly(styrene) have been used to measure structure and diffusion dynamics. Diffusion entry was measured by adding a fluorescence quencher, ethyl oxalate, which cannot enter the polyethylene matrix.

The data indicated the presence of segregated domains with extensive interactions at the domain interfaces. Using pyrene-3-

carboxyaldehyde the interaction between sodium dodecyl sulfate and acrylic copolymers has been found to occur at concentrations smaller than that observed for the pure surfactant in water. An increase in polymer concentration lowers the critical aggregation concentration. The cyclization of poly(ethylene oxide) endtagged with pyrene groups has been monitored, with excimer formation being described by a two state model. At temperatures below 30°C a pyrene dimer is formed, with activation energies close to those for viscous flow activation. The difference between the enthalpy and the binding energy of the excimer in both the globule and coil states reflects the variation of polymer conformations between the cyclized and noncyclized chains.

In LC pyrene-labelled cholesterol copolymers, each pyrene molecule is apparently incorporated into the LC domain with restricted access, while in poly(styrene) latexes small hydrophobic pores have been identified through monitoring the extent of pyrene penetration. Via the same probe, it has been shown that thermally reversible phase separation of poly(isopropylacrylamide) is due to coil collapse. Thus, trapping and release of the probe within the polymer coils will give rise to different spectra and emission lifetimes.

The folding of pyrene labeled poly(*N*,*N*-dimethylacrylamide) has been described by a blob model. This technique provides valuable details on chain dynamics in solution, including the rate of segmental encounter. Thus, the folding of a polymer chain can be handled by compartmentalizing the polymer coil into subdomains, typical of that used for proteins. Inter-chain interactions have also been measured in an ethylene-propylene random copolymer tagged with 1-pyrenebutanoic acid hydrazide. Using the blob model, 70% of the pyrene groups were found to be closely located, while in an apolar solvent over 87% of the groups are involved in aggregate formation held together by succinic anhydride groups. Poly(*N*-(1-pyrenylmethyl)methacrylate)) behaves in the same way. Hydrophobically modified copolymers of *N*-isopropyl-acrylamide and*N*-?-valineacrylamide undergo pH dependent phase separation on heating, while in micellar media the quenching of probes by cetyl pyridinium chloride follows the order pyrene > perylene > fluoranthene.

The interactions between cationic non-phospholipid liposomes and pyrene-labelled hydrophobically modified sodium poly(2-2-(acrylamido)-2-methylpropane sulfonates) indicates that the polyanions bind to the cationic liposomes.

The reversibility of this process was also demonstrated. Energy transfer and excimer formation of a general interest have been widespread. Sulfonated polystyrene has been shown to give rise to monomer, excimer and dimer emission bands, while the motion of end-capped anthryl groups in polystyrene provides a useful model for predicting the orientational auto-correlation function. Bis(10-phenyl-9-dianthryloxy)methane excimer formation is sensitive to the viscosity of poly(dimethyl) siloxanes, while increasing the styrene content in poly(styrene) by irradiation enhances excimer formation. Polymers with 100% donor-acceptor norbornadiene groups undergo effective isomerism, the effect being enhanced in the presence of a benzophenone triplet sensitizer.

Fluorescein and phenolphthalein have been shown to differ markedly in their photoelectric properties, while the donor-acceptor properties of naphthalene and pyrene tagged to poly(ethylene glycol) units on a poly(styrene) chain are affected by external stimuli. In micellar media with increasing salt addition, the chains coil, thus reducing the separation distance between the optical tags.

Fullerenes are quenched by electron transfer from polyimides with a strong dependence upon the orientation of the chain. Here*m*-substitution with a high ionization potential exhibits the lowest quenching rate. Micro- photoluminescence has been used to measure excitonic luminescence in a single conjugated polymer chain. The data indicates inhomogeneous broadening, very long chains and long-range rapid exciton transfer (< 100 ps).

Interface distances between films of block acrylic-styrene copolymers have been measured by energy transfer between anthryl and phenanthryl labeled junctions, while the fluorescence quantum yield of 2-*N*-piperidino-5-(2,2'- dicyanovinyl)thiophene is enhanced from 0.0004 in hexane at room temperature to unity in EPA-glass at 77K. The analysis of delayed fluorescence from various conjugated polymers at different temperatures indicates that the emission occurs within a time frame of micro- to milliseconds at 80 K, while the phosphorescence on average occurs over a one second period. This suggests that delayed fluorescence is actually caused primarily by the recombination of geminate electron-hole pairs rather than triplet-triplet annihilation.

The effect of an external electrical field in the experiments tends to confirm this deduction. FRET in folded polymer chains has also been under the 'microscope'. Simulations suggest that FRET can act as a good marker of polymer folding only when the Forster radius (Rf) is smaller than the root-mean-square radius (R0) of the polymer.

It is suggested that it may well be important to employ more than one donor-acceptor pair in experiments in order to probe the entire dynamics of chain folding.

A new method for the synthesis of polymers with bifluorophores has been developed. In this case it was concluded that nonradiation energy transfer is not feasible if the distance between donor and acceptor exceeds 10 nm. In another analysis of T-T annihilation in conjugated polymers, it was also concluded that this process was effective up to 3% of the total luminescence.

Super-quenching has been observed in poly derivatives tagged with cyanine dyes on each repeat unit. Strong aggregate formation was evident in these polymers, with quenching rates equivalent to or greater than those normally observed with polyelectrolytes. Different types of complexations have been observed in carbazole containing polymers, and FRET dependence upon temperature has been investigated.

Activation energies have been measured for main chain segmental motions in anthracene tagged PVK's. Relaxation times were measured through fluorescence anisotropy decay, giving energies greater than those measured for many other polymer systems. It is claimed thatPVKhas a larger critical molecular weight factor and that the bulky side-chains suppress the conformational transition of the polymer backbone.

Excimer formation in PVK copolymers indicates that nearest neighbour interactions are dominant, while acrylic copolymers of PVK give stronger emission than the monomers. The twisting of polymer chains in oligo[2,5-bis(hexadecyloxy-1,4-phenylene]s reduces excimer formation by preventing stacking, whereas hydrophobic interactions due to long aliphatic chains cause excimer formation. Spatial structures have been measured in bifluorophore azomethine polymers, and carrier generation and transport examined in general.

Polymer tagging is useful as a macromolecular probe. Naphthalene has been tagged to cyclodextrin for potential light harvesting, while anthracene has been tagged to PMA to determine the efficiency of photodimerization. In anthracene labeled PMMA relaxation times have been determined, whereas spectral broading can be used to determine the size of a polymer molecule.

Poly(imidoaryl ethers) have been tagged with bulky *N*-phenyl-naphthalimides and characterized spectroscopically, as have novel 3,4,6-triphenyl-a-pyrone fluorophores where the phenyl group is fixed in a specific orientation for fluorescence. Other characterizable polymers include poly(2-ethynyl-pyridinium bromide) with propargyl side chains, naphthalene tagged poly(butenoate), dye labeled poly(DL-lactic acid) and proteins, cyanine dyes on nucleic acids, polyphosphazines with carbazolyl side chains, coumarin dendrimers and anthracene labelled epoxy resins.

Polystyrene has been shown to give high yields of dimers when formed with dibromomethane, and, when tagged with fullerene, the emission is blue shifted owing to a loss of conjugation. Poly(acrylics) have also been tagged with naphthalimide dyes and spectrally characterized.

The fluorescence of poly(imides) has been controlled through the tagging of perylene and quinioxaline units, with little aggregation below 2.82 mol-%. The polymers also exhibit complete quenching due to efficient electron-transfer processes. Poly(*N*-isopropylacrylamide) has been tageed with tetraphenyl-porphyrin groups and found to experience a thermally reversible phase transition upon heating. Fluorene has been tagged to polysiloxanes and found to exhibit higher rotational and translational mobilities than for conventional siloxane polymers.

Poly(dipropylsilylenes) have been made by vapour deposition and characterized, and dansyl groups on a poly(dimethylsiloxane) have been used to monitor mobility at junction interfaces in liquids and molten polymers.

Here the solvent is concluded to play a dramatic role in tuning the local microenvironment that surrounds a polymeric junction point. The biodegradability of poly(amide-anhydrides) can be measured via visibly fluorescent tags, and the spectroscopic properties of vinylcaprolactam copolymers with vinylpyrrolidone have been determined in the presence of cations.

Polymers doped with rare earth metal ions continue to be explored for enhanced optical properties. Poly(acrylic acid)-Eu(III)-dibenzoyl methane exhibits high fluorescence quantum yields, as do ternary Te(III) complexes in styrene-acrylics. Such systems can have useful applications, in particular Ru(II) complexes in polydimethylsiloxanes have been used as 'pressure-sensitive paints'. Water molecules trapped in Eu(II) complexes in poly(methacrylic acid) have been estimated, whereas the stability of Tb(II) ions in *N*-alkylmethacrylamide copolymers depends upon the structural organization of the macromolecules.

Alkylpyrrole groups influence the emission quantum yields of polypyrrole Ru(III) complexes, with decay rates that are affected by extents of crosslinking. The use of methylviologen for redox processes is still feasible in crosslinked systems. Stern-Volmer quenching measurements by T_1(I) ions have been undertaken on fluorescent dyes doped in sodium dodecyl sulfate, and energy transfer assessed between Ru(II) and Os(II) bipyridyl compexes attached to the ends of a deca(oxoethylene)-33 atom chain.

N-vinylcarbazole and *N*-vinylpyridine have been complexed with Ir, each showing quenching of the excimer sites. Inorganic phosphors have been doped in polymers to produce light emitting caps for bottles, while lanthanide complexes in 3(trimethoxysilyl)propyl methacrylate give highly stable luminescent matrices. In a similar way, Eu(III) complexes in methacrylate copolymers bound onto silica gel provide a useful medium for intensifying the luminescence output. Fluorescence probe studies in micellar matrices attract much interest.

Micellar size and structural characterization of self-assembled systems have been covered in some depth. Copolymers of poly(isoprene-b-ferrocenylphenylphosphine) form starlike spherical micelles in *n*-hexane with a dense polyferrocene core. The collapse of poly(*N*-isopropyl acrylamide) in water is dependent upon the amount of grafted poly(ethylene oxide).

A pyran probe was found to localize in the hydrophobic parts of the copolymer chains and exhibited properties that were highly dependent upon the conformation of the chains during garfting. The effects of poly(ethylene oxide) in triblock copolymers have similarly been studied using a pyrene derivative as a probe, including in one study the amount of water trapped in micelles.

Silicone surfactants with hydrophilic amine groups have also been grafted with poly(ethylene oxide) chains and found to exhibit aggregates in aqueous solutions. For related poly(oxyalkylene-substituted siloxane) surfactants in the presence of methyl orange and pyrene probes, micelles are formed with hydrophobic interiors with CMC values that shifted to higher values in mixtures of co-solvents such as ethanol. Interactions between butyl vinyl ether-sodium styrene sulfonate copolymers and oppositely charged ionic surfactants show premicelle formation owing to electrostatic attractions where free probes (pyrene) will migrate from the aqueous to the new hydrophobic sites.

A series of novel polyazo electrolytes have been prepared, which exhibited interesting photochromic effects depending upon the functionalization and coating processes. Also, the continuous addition of surfactant destroys the excimer sites. Premicellar formation has also been observed using a pyrene probe for a series of Gemini surfactants, while poly(*N*-acetylethyleneimine) cosurfactants were found to be included in alkyl methacrylate copolymers made by emulsion polymerization.

Twin-tailed hydrophobic acrylic monomers have been prepared by micellar methods and, using a dansyl labelled probe, the relative interactive strengths of different surfactants were found to be in the order CTAB > Triton-X > SDS. Binding processes in CTAB micelles have also been investigated using dansyl labelling.

Local mobility was confined to the alkyl tails of the surfactant, with polyelectrolyte-surfactant complexes exhibiting an ordered lamella structure. The use of pyrene fluorescence as a probe has shown that KCl decreases the CMC of Triton-X-100 as well as increasing the aggregation number and degree of hydration.

The latter is due to water trapped in the micelle, although there were no changes in the micellar micropolarity owing to solubilization of the probe associated with micellar growth. Certainly, electrolyte addition enhanced the microviscosity, which is consistent with micellar hydration.

Cyanine dyes have been shown to interact with DNA at low concentrations, while similar dye chrompohores exhibit a much narrower absorption band in poly(vinyl alcohol) than they do in poly(vinyl sulfate). Self-assemblies of fluorescent dyes such as rhodamine B have been demonstrated on zirconium phosphate, where binding reactions give rise to significant enhancement in quantum yields. Energy transfer has been observed not only in these dyes but also in new terpolymers. A sugar cationic surfactant containing a fluorescent coumarin group exhibits a certain chirality, which causes bending forces, resulting in the formation of helixes and tubules, while a series of carbazole amphiphilic copolymers exhibits emission characteristics that are dependent upon the environment, and several binding modes were ascertained for diaminostilbene and polyions.

Hydroxyethylcellulose with cationic hydrophobic sidechains form micelle-like aggregates, with hydrogen-bonding playing amajor role in the interactions. Amino derivatives of 7-nitrobenz-2-oxa-1,3-diazole have been found to be sensitive to polarity and microviscosity changes for amphiphilic copolymers, whereas poly(methacrylic acid) and poly(vinylpyrrolidone) undergo complexation via a connected cluster model. Fluorescent sensors have been developed which can complex with the ammonium groups of starch, while end-capped poly(sodium-2-acrylamido-2-methyl- propane-sulfonates) undergo multi-molecular aggregation in salt water solutions. In the presence of CTAB and a pyrene salt probe, the latter is excluded from any binding. An ionic derivative of pyrene has been bound to a polyelectrolyte, while other novel probes have been developed.

Copolymers of acrylic acid and*N*-dodecylmethacrylamide form micelle-like multipolymer aggregates, while twin-tailed terpolymers reveal viscoeleastic properties consistent with a simple Maxwell model. For a series of poly(allylammonium) chloride salts with differing alkyl side-chains, hydrophobic domains depend upon the length of the alkyl sidechain. Using a pyrene derivative probe, these changes are assigned to aggregation arising from interchain interactions. In fact, polymers with dodecyl side chains give hydrophobic contents where aggregates are independent of the polymer concentration.

In the case of *N*-vinylcarbazole labeled methacrylic acid-acenphthylene copolymers, a hypercoiled conformation at low pH makes the fluorescence primarly that of the acenaphthylene excimer. At high pH, however, the copolymers adopted an extended coil conformation, thus decreasing non-radiative energy transfer.

Surfactants, micelle formation and aggregation have been critically assessed in some detail. General polymer luminescence studies are numerous. Crosslinked polyethylene is shown to give emission due to recombination centres, while the luminescence from polypropylene composities depends upon the composite. Novel conjugated poly(1,2-diethynyl-1,1,2,2-tetrmethyldisilanes) are moderately luminescent, while novel luminescent binaphthyl polymers have also been made. Polyanhydrides give fluorescence only when a methane group is in the repeat unit, as do polyamides with a repeat unit of ethidium bromide.

Fluorescent dyes have been used to measure polymer melt temperatures during extrusion, while some polyethers give a weak blue fluorescence. Poly(ethyle-neterephthalate) gives luminescence due monomeric components, and poly(pro-pylene) doped with 9,10-diphenylanthacene enables the detection of a true infectious profile on thermal oxidation. Polyaniline exhibits fluorescence, which is maintained in a vacuum but removed and red-shifted upon acidification. This also influences the conductivity of the polymer. Electron beam irradiation affects the luminescence from poly(vinyl toluene), while the emission properties of phenyl-substituted siloxanes does not depend upon the phenyl groups.

Ultrafast intra-chain vibrational relaxation processes have been measured in poly(fluorene) films. Whereas composites of the same polymers possess a hierarchy of phases, the composition affects photovoltaic properties. A fibre-optic cure sensor based on evanescent-wave fluorescence spectroscopy has been designed to probe the inter-phase region of GRP composites and plasma prepared organo-siloxanes give emissions associated with changes from linear to amorphous structures.

The same process applied to poly(propylene) gives rise to three spectral components. These are two fast-decay processes, due to radiative de-excitation levels, and a slower carrier-recombination process. The luminescence of the same polymer also depends upon its crystal structure while that from a series of poly(silanes) depends very much on polymer structure and conformation.

Stern-Volmer quenching has been measured in poly(*N*-*p*-phenoxyphenyl)methacrylamide, and aggregation affects the luminescence from polymers from 2,5-dioctyl-*p*-distyryl-benzene, perylenebisimide-polytetrahydrofurans and sulfonated poly(butyl-methacrylate)-polystyrene.

Continuing studies of general optical properties include the luminescence of copoly[arylene-1,2-dioxy-oligodimethylsiloxanylenes], pressure sensitive paints, nanoparticle acrylics, fullerenes, polystyrene particles, polyfluorene, poly(4-vinylbenzyl-thiocyanate), polystyrene films, PMMA, stilbenes, complexes of *N*,*N*-dimethylacrylamide and PMA, poly(diallyldiethyl-ammonium chloride), polymers of 9,10-dibromoanthracene and 1,4-diethynylbenzene, poly(Schiff 's bases), poly (heteroarylenes) and poly(di-*n*hexylsilanes).

The luminescence of poly(ethylene naphthalate) is associated with the presence of degradation products, as is the case with other materials. Luminescence has been used as a molecular probe for identifying and pin-pointing treeing sites in PMMA. The lasing properties of diamine-xylylene copolymers have been examined, while fluorene-based fluoro-polymers emit blue light except those with a perfluoro-octyl group.

Photodegradation and Photooxidation Processes in Polymers

The photodegradation of polymer materials continues to attract interest, although as in the last few years activity has declined. Thermal degradation and stabilization processes rank much more highly in that particular field. Laser ablation processes continues to grow in interest, with applications in electronics. A useful review on the subject has appeared, giving extensive coverage of techniques and applications.

Polyolefins

Polyolefin photooxidation processes is the most actively studied area. Depth profiles have been determined in tropical climates for toughened polypropylene, while other work has covered resistance to discolouration.

Iron complexes continue to be investigated as effective pro-oxidants for polyolefin photodegradation., Iron is a reknowned catalyst for olefin oxidation, and the problem here has been to optimize processing stability while at the same time producing a formulation which will sensitize photooxidation of the matrix within a given time period. Iron complexes have also been examined in conjunction with starch for applications with poly(ethylene)., Chemical and physical changes during poly(ethylene) photooxidation continue to be measured, while other work has also dealt with the effects of doping with rubber and manganese stearate.

Crosslinking in poly(propylene) during irradiation has been measured, similarly for PP foam material used in snack boxes and for flame retardency. Ethylene-carbon monoxide copolymers have also been investigated for structural and chemical changes during irradiation. Through mass measurements of oxygenated products, it is interesting to note that while peroxy radical termination is evident in photooxidation it is absent in thermal oxidation. Polymer-alkene-fullerene radical products have also been identified during the irradiation of doped polymer materials, using benzophenone as a photosensitizer.

Hydroperoxide sites have been enhanced on the surface of polyolefins by grafting with poly(styrene), and poly(ethylene) made via the spherilene technology becomes highly unstable on irradiation, with rapid growth in carbonyl and hydroperoxide groups. Changes in the properties of poly(ethylene) have also been measured under the influence of Sahara winds, while secondary usage of poly(ethylene) gives generally an inferior product with limited usage.

Polystyrenes and Polyacrylics

Radical synergy has ben observed in poly(styrene) doped with three initiators. In fluorinated methacrylate formulations, the overall light stability is controlled by the butyl- methacrylate component, while irradiation of PMMA can cause chain scissions, giving rise to mechanical failure. With far UV exposure there is also a tendency towards depolymerization. In other work, methacrylates have been shown to be more unstable than acrylates, owing to associated ester chain scissions in the former case.

Polyesters

The surface degradation of poly(ethylene terephthalate) has been examined by X-ray methods.

Polyamides and Polyimides

Polyimides with *N*-carbonyloxyimide groups undergo chain scission through N-O cleavage as well as breaking of the cyclobutane ring.

Silicone Polymers

The fluorescence from poly(methylcyclohexylsilanes) decreases and blue shifts upon irradiation, with incorporation of siloxane groups in the Si–Si bonds. Poly(dimethyldisiloxy) materials with acetophenone groups attached display high instability, showing a 50% reduction in molecular weight over a short period of irradiation. Spectroscopic analysis of poly(silanes) shows that irradiation at longer wavelengths tends to scission the longest segments on polymer chains, while short wavelengths induce degradation of all chain lengths.

Polyurethanes and Rubbers

Urethane coatings based on Castor oil develop strong yellowing on UV exposure, while the same is true of aromatic based urethanes rather than aliphatic based systems.

The polyether segments are attacked first, forming ester and other volatile products. There is a loss in free-volume and holes during poly(urethane) photodegradation, coupled with a high degree of crosslinking., Irradiation of MDI-based urethanes in the presence of nitrogen dioxide causes nitration of the aromatic rings and scission of the methyl substituents followed by crosslinking, while ESR has been used to identify radicals formed during irradiation of poly(urethanes) under different light sources.

The melt rheology and post irradiation of poly(octenamer) have been examined, and in styrene-isoprene-styrene blends the isoprene matrix is the most unstable. Initially, chain scission occurs followed later by crosslinking. Spatial heterogeneous degradation processes have been observed in ABS terpolymer, whereas irradiation of natural rubber leads to the formation of liquid rubber as a product via the usual hydroperoxidation reactions.

Poly(Vinyl Halides

PVC used in the construction business has been exposed to various external environments, and it was shown that whilst stress changes slightly there is no change in hardness. Chemical changes were, however, apparent by FTIR, with ozone being the most detrimental external agent.

Photoablation of Polymers

Laser ablation is one of the most widely studied areas in the field of polymer degradation. The yield of silyl radicals during KrF laser ablation of poly(silylenes) has been found to be dependent upon the backbone structure of the polymer, whereas the yield of silylenes was found to be proportional to the number of linear Si units in the backbone chains. A fluffy nanostructured powder of poly(hydridomethy-ldisiloxane) has been prepared by laser CVD of 1,3-dimethyldisiloxane, as have other organyltrimethylsilane polymers. Temperature effects on the deposition of laser ablated hexaphenyldisilane have also been investigated, with a view to optimizing film properties.

In polyamides and polyimides, the high-fluence implantation of boron ions, mainly via electron stopping, has been found to result in the formation of electron rich p-bonded carbon clusters. In contrast, heavier phosphorus ions, for example, deposit energy predominantly by nuclear collisions to give a lower concentration of p-radicals and a less carbonized surface. Irradiation ablation of poly(methacrylates) with 146 nm light has been found to be dependent upon the structure of the ester groups, while for poly(styrene) no thickness decrease was observed, with some derivatives exhibiting crosslinking. The luminescence from poly (propylene) decreases with laser ablation time, while PVDC has been found to be more sensitive to ablation than PVC.

Using poly(ethyleneter- ephthalate) as a model substrate, it has been shown that new functionalities can be created in microchannels by laser ablation. Nanoscopic aggregation has been studied in laser irradiated amphiphilic random copolymer environments. Structure-property relationships have been established for a series of polymer materials at low irradiation fluences. Polymers with photochemically active groups were found to have the lowest threshold for ablation. The thermal stability of the polymers was not an important parameter. In this way, polymers could be designed for specific ablation processes in order to develop microoptical components. A time-resolved ultramicroscope has been developed in order to monitor the process of pulsed-laser induced melting of 9,10-dicyanoanthracene microcrystals on poly(ethyl methacrylate) films. This instrument is claimed to be able to detect sub-micrometer particles on a time resolution of 400 ps.

Crystals with a size of 10 microns were found to be converted into hot liquid droplets, of which some 20% had evaporated. Mixtures of 3,4,9,10-perylenetetracarboxylc dianhydride with Co powder have been found to give fragments upon laser ablation at 308 nm that no longer have carboxylic anhydride groups. IR analysis shows on poly(perinaphthalene) is deposited, giving highly conductive films. Collagen and keratin nave been successfully deposited, with no change in their chemical structure but an increase in their random-coil domans.

A polyurethane film has been successfully etched away. with no debris, using 248 nm laser light, while collagen gels have been found to exhibit smaller threshold energies than in their dry state. Here the latent heat of water evaporation absorbs most of the laser energy.

Natural Polymers

The discolouration and protection of wood and paper continues to attract some interest. The extensive nature of the subject has been made evident, including modification treatments. Hydroxyl radicals are responsible for the degradation of glycosidic linkages, giving rise to a series of alsonic acids and aldoses.

Photochemical bleaching of wood pulp has been found to be much more effective, while acetylation of wood reduces postphotoyellowing. This process inhibits the photolysis of phenoxyl ether groups giving rise to phenoxy radicals. Butyrylation has been reported to be even more effective than acetylation, giving rise to a significant reduction in quinoid products. Modification experiments, utilizing grafting with methacrylic anhydride, also inhibit yellowing. The interesting feature of this study, however, was the observation that degradation still continues and that yellowing is a separate process. Photoyellowing of wood substrates and products has been found to be impaired by 2-hydroxyaromatic stabilizers, while glutathione is a photosensitizer for collagen.

Miscellaneous Polymers

Polymers and coatings of various compositions have been investigated and mechanisms reviewed. A dual photoo- xidation process has been proposed for trimethy- lcyclohexane-polycarbonate, where under short wavelengths a phot-Fries rearrangement is operative as well as chain scission. This polymer was also found to be more photoreactive than normal polycarbonate, owing to the higher incidence of reactive tertiary carbons in the trimethy-lcyclohexylidene structure. For organic coatings, hardness, *Tg* and chemical changes by FTIR have all been inter-related during UV ageing.

The service lifetime of acrylic-melamine coatings has been measured in a series of ageing apparatus and accelerated weathering tests evaluated for a series of epoxy-acylurethane paints. The photooxidative stability of a series of EB- and UV-cured dialkyl and heterocyclic amine terminated acrylic resins has been examined. UV-cured coatings have been found to be more unstable than those EB-cured, with a strong structural dependence upon the nature of the terminal amine functionality. Thus, dialkylamines were found to be more prone to oxidation than alkanolamines, cycloaliphatic amines and heterocyclic amines.

Yellowing is associated with facile oxidation of the amine group by hydrogen atom abstraction, to give unsaturated carbonyl products. Photoirradiated radical yields in acrylic and alkyd paint films have been compared and found to correlate with stability. Fluorinated acrylates have also been studied, as have hot-melt adhesives, UV-cured clear coats and epoxy resins. For a series of poly(benzoxazines), photoo- xidative stability has been found to depend not only upon the nature of the phenolic substitution but also upon the type of amine functionality present. On photooxidation, benzoquinones are formed, giving rise to discolouration.

Hydrogen bonding between the phenolic and amine functionalities was also important. Photodegradation depth profiles have been determined in thick section polycarbonates, while the quantum efficiency of luminescence form PPV's after a few minutes irradiation decreases significantly.

Poly(4'-ethoxyacrylophenone) undergoes random chain scission with a higher quantum yield in solution than in the solid state. Ethyl radicals were formed in high yield due to O-alkyl bond scission, coupled with yellowing due to quinonoid entities. Crosslinking occurs later with a consequent reduction in chain scission due to restricted migration of the free radicals.

The fire-retardency of nanocomposites is important, while poly(ethylene glycol) hydrogels are made photoreactive by adding cinnamate groups Super macromolecules are formed in poly(vinyl alcohol) on irradiation in the presence of Fenton's reagent, while high levels of formates are generated during the photooxidation of poly(ethylene oxide), owing to the predominance of peroxy radical termination processes. For polycarbonate-polyurethane structures, photolysis occurs at the C–H bond adjacent to the urethane group, causing rapid chain scission, whereas the microstructure of poly(octenamer) is important in photoooxidation reactions of the polymer. Thus, whilst rapid crosslinking was observed there were no corresponding changes in FTIR functionalities.

Hydration effects on the stability of polyacrylics has been determined, and poly(naphthalimides) modified. Other studies of interest include surface anisotropy due to bond breakage, use of DSC methods for hydroperoxide analysis, photohydrolytic compatibility of polyethylene and polycaprolactone and new polymeric sensitizers.

Photostabilization of Polymers

The photostabilization of polymer materials for outdoor applications and extended life continues to be an industrial battleground. Consequently, many of the publications are commercial in nature and centre on effects rather than scientific processes.

Numerous articles of interest include the surface protection of poly(propylene), new cyanoethylacetophenone derivatives for poly(ethylene) modification, stabilization of rotomolded parts, hindered amine stabilizers (HAS) for glass-reinforced nylon-6, novel additives for poly(propylene) fibres, additives for packaging products, novel cycloaliphatic amines for HAS, absorbers for elastomers, computer-aided design for packages, HAS and absorbers for plastics and coatings in general, stabilization of rigid PVC and wood composites, stability of white pigmented ABS, new blends for urethanes, stabilization of natural woods, flame retardant ABS, HAS for PVC, novel aryl and heteroaryl methacrylate derivatives, HAS for styrenics, colour interactions of HAS and antioxidants, HAAS for automotive paints, absorbers for fabrics, cycloalkylated products for isoprenes, electronic properties of absorbers, hexaazocyclanes and photo-Fries products for polyesters and enhanced packages for fibres.

In the polyolefin world, the development of metallocene catalysts has attracted much interest, and in this regard stabilization processes have been examined and found to be the same as for conventional polymer grades. During the ageing of polyolefins with a polymeric HAS system, fragments of stabilizers have been identified, with extreme conditions in water and low pH being the most detrimental. HAS grafted to the surface of polyolefins has been claimed to give the best surface protection, while concentration changes in HAS has shown that the original amine decreases slowly, whereas the nitroxyl radicals increase and then after a certain period decrease. It was also shown that, whilst the amine remains uniform throughout the polymer material, the nitroxyl radicals exhibit non-uniformity. Mixtures of HAS structures have been shown to operate synergistically in the light stabilization of polyolefins, as have other stabilizers based on hydroxyphenyltriazine with oxanilide or benzophenone. The incorporation of HAS into urethane coatings has shown that the stabilizer increases the number of urea bridges at the expense of unreacted isocyanate, giving rise to a more photostable material.

The loss of UV absorbers has been measured in clear-coats during UV ageing,, while for UV-cured acrylates HAS are effective radical scavengers. PVC plates have been made more light-stable by applying a UV curable acrylic coating containing both a HAS and an absorber, while, in the case of wood pulp, the use of a hydroxyphenylbenzotriazole absorber and a nitroxyl radical has been found to be effective in inhibiting the photolysis of the aryl phenyl ether groups. Tertiary phosphine complexes of nickel are shown to be highly effective light stabilizers for polystyrene only when methyl and allyl substituents are present. Finally, UV absorbers give some UV protection to poly(ethylene terephthalate).

Photochemistry of Dyed and Pigmented Polymers

With regard to dyes and pigments, there continues to be moderate interest, with much of the activity centred on the catalytic effects of titanium dioxide. PVC profiles under northern climatic exposures in wet conditions are known in some cases to induce an unacceptable 'pinking' phenomenon. Here certain grades of titanium dioxide can induce or alter the balance of chemical reactions to form polyene sequences. Photoactivity of titanium dioxide pigments has also been measured via low temperature luminescence, with multiphasic materials showing no evidence for carrier segregation. This pigment has been used to destroy formaldehyde on carbonized wood surfaces as well as providing ecological photocatalytic coating surfaces., Paper containing photocatalytic titanium dioxide pigment decomposes ammonia gas as well as being useful for the destruction of cyanide ions.

Antibacterial photocatalytic coatings have also been developed, as have novel processes for photobleaching pul. In the photocatalytic decomposition of a reactive red dye, effective decomposition occurred only under visible light exposure, whereas a zinc oxide pigment doped with Ag was highly effective in decolourizing a blue dye.

Aluminasupported zinc oxide has also been shown to be highly effective in photobleaching a green acid dye with an optimum at a pHof 4. The weathering behaviour of pigmented polymers. One interesting observation is that coating of filler particles with a surfactant ehances their ability to stabilize a polymer. The natural dye Carthamin has been effectively stabilized using a singlet oxygen nickel complex quencher, as have a series of triphenylmethane dyes used for carbonless copying papers. Nickel hydroxypheny lbenzotriazolesulfonate absorbers are also useful in this regard, as are ester type benzotriazole stabilizers.

Cellophane has been shown to be a good model for the photofading of dyes on paper and cotton, whilst moisture is important in the photofading of vat dyes in polyesters, with finishing treatments also having a potentially detrimental effect. Reactive bonded dyes to cotton are more light stable, while the products of dye fading can also be important in reaction kinetics. In multilayered coatings of PVC and fluorinated polymers, the photostability of fluorescent dyes has been found to be dependent upon the self-absorption properties of the dyes. Blends of polymers are also shown to be more effective.

The presence of molecular oxygen inhibits the fading of 4-diethylaminoazobenzene dyes in solutions and solid state, while cyanine dyes with a positive charge are more photostable than corresponding mercocyanines. Rose Bengal has been used to accelerate the photofading of triphenodioxadine dyes in cellulosic films.

Chapter 5

Environmental Chemistry Basics

Environmental chemistry is the scientific study of the chemical and biochemical phenomena that occur in natural places. It should not be confused with green chemistry, which seeks to reduce potential pollution at its source. It can be defined as the study of the sources, reactions, transport, effects, and fates of chemical species in the air, soil, and water environments; and the effect of human activity on these. Environmental chemistry is an interdisciplinary science that includes atmospheric, aquatic and soil chemistry, as well as heavily relying on analytical chemistry and being related to environmental and other areas of science.

Environmental chemistry is that branch of one,which deals with the study of chemical and biochemical phenomena that occur in natural places like air,soil,water.

Environmental chemistry involves first understanding how the uncontaminated environment works, which chemicals in what concentrations are present naturally, and with what effects. Without this it would be impossible to accurately study the effects humans have on the environment through the release of chemicals.

Environmental chemists draw on a range of concepts from chemistry and various environmental sciences to assist in their study of what is happening to a chemical species in the environment. Important general concepts from chemistry include understanding chemical reactions and equations, solutions, units, sampling, and analytical techniques.

Contamination

A contaminant is a substance present in nature due to human activity, that would not otherwise be there The term contaminant is often used interchangeably with *pollutant*, which is a substance that has a detrimental impact on the environment it is in. Whilst a contaminant is sometimes defined as a substance present in the environment as a result of human activity, but without harmful effects, it is sometimes the case that toxic or harmful effects from contamination only become apparent at a later date.

The “medium” (e.g. soil) or organism (e.g. fish) affected by the pollutant or contaminant is called a *receptor*, whilst a *sink* is a chemical medium or species that retains and interacts with the pollutant.

Environmental Indicators

Chemical measures of water quality include dissolved oxygen (DO), chemical oxygen demand (COD), biochemical oxygen demand (BOD), total dissolved solids (TDS), pH, nutrients nitrates and phosphorus), heavy metals (including copper, zinc, cadmium, lead and mercury), and pesticides.

Applications

Environmental chemistry is used by the Environment Agency (in England and Wales), the Environmental Protection Agency (in the United States) the Association of Public Analysts, and other environmental agencies and research bodies around the world to detect and identify the nature and source of pollutants. These can include:

- Heavy metal contamination of land by industry. These can then be transported into water bodies and be taken up by living organisms.
- Nutrients leaching from agricultural land into water courses, which can lead to algal blooms and eutrophication.
- Urban runoff of pollutants washing off impervious surfaces (roads, parking lots, and rooftops) during rain storms. Typical pollutants include gasoline, motor oil and other hydrocarbon compounds, metals, nutrients and sediment (soil).

Methods

Quantitative chemical analysis is a key part of environmental chemistry, since it provides the data that frame most environmental studies.

Common analytical techniques used for quantitative determinations in environmental chemistry include classical wet chemistry, such as gravimetric, titrimetric and electrochemical methods. More sophisticated approaches are used in the determination of trace metals and organic compounds. Metals are commonly measured by atomic spectroscopy and mass spectrometry: Atomic Absorption Spectrophotometry (AA) and Inductively Coupled Plasma Atomic Emission (ICP-AES) or Inductively Coupled Plasma Mass Spectrometric (ICP-MS) techniques. Organic compounds are commonly measured also using mass spectrometric methods, such as Gas chromatography-mass spectrometry (GC/MS) and Liquid chromatography-mass spectrometry (LC/MS). Non-MS methods using GCs and LCs having universal or specific detectors are still staples in the arsenal of available analytical tools.

Other parameters often measured in environmental chemistry are radiochemicals. These are pollutants which emit radioactive materials, such as alpha and beta particles, posing danger to human health and the environment. Particle counters and Scintillation counters are most commonly used for these measurements. Bioassays and immunoassays are utilized for toxicity evaluations of chemical effects on various organisms.

Published Analytical Methods

Peer-reviewed test methods have been published by government agencies and private research organizations. Approved published methods must be used when testing to demonstrate compliance with regulatory requirements.

Freshwater Environmental Quality Parameters

Freshwater environmental quality parameters are the natural and man-made chemical, biological and microbiological characteristics of rivers, lakes and ground-waters, the ways they are measured and the ways that they change. The values or concentrations attributed to such parameters can be used to describe the pollution status of an environment, its biotic status or to predict the likelihood or otherwise of a particular organisms being present. Monitoring of environmental quality parameters is a key activity in managing the environment, restoring polluted environments and anticipating the effects of man-made changes on the environment.

Characterisation

The first step in understanding the chemistry of freshwaters is to take samples and analyse them for the chemical constituents that are of interest.

Sampling

Freshwaters are surprisingly difficult to sample because they are rarely homogeneous and their quality varies during the day and during the year. In addition the most representative sampling locations are often at a distance from the shore or bank increasing the logistic complexity.

Rivers

Filling a clean bottle with river water is a very simple task, but a single sample is only representative of that point along the river the sample was taken from and at that point in time. Understanding the chemistry of a whole river, or even a significant tributary requires, prior investigative work to understand how homogeneous or mixed the flow is and to determine if the quality changes during the course of a day and during the course of a year. Almost all natural rivers will have very significant pattern of change through the day and through the seasons. Many rivers also have a very large flow that is unseen. This flows through underlying gravel and sand layers and is called the hyporheic zone How much mixing there is between the hyporheic zone and the water in the open channel will depend on a variety of factors some of which relate to flows leaving aquifers which may have been storing water for many years.

Ground-waters

Ground waters by their very nature are often very difficult to access to take a sample. As a consequence the majority of ground-water data comes from samples taken from springs, wells, water supply bore-holes and in natural caves. In recent decades as the need to understand ground water dynamics has increased, an increasing number or monitoring bore-holes have been drilled into aquifers

Lakes

Lakes and ponds can be very large and support a complex eco-system in which environmental parameters vary widely in all three physical dimensions and with time. Large lakes in the temperate zone often stratify in the warmer months into a warmer upper layers rich in oxygen and a colder lower layer with low oxygen levels. In the autumn, falling temperatures and occasional high winds result in the mixing of the two layers into a more homogeneous whole. When stratification occurs it not affects oxygen levels but also many related parameters such as iron, phosphate and manganese which are all changed in their chemical form by change in the redox potential of the environment.

Lakes also receive waters, often from many different sources with varying qualities. Solids from stream inputs will typically settle near the mouth of the stream and depending on a variety of factors the incoming water may float over the surface of the lake, sink beneath the surface or rapidly mix with the lake water. All of these phenomena can skew the results of any environmental monitoring unless the process are well understood.

Mixing Zones

Where two rivers meet at a confluence there exists a mixing zone. A mixing zone may be very large and extend for many miles as in the case of the Mississippi and Missouri rivers in the United States and the River Clwyd and River Elwy in North Wales. In a mixing zone water chemistry may be very variable and can be difficult to predict. The chemical interactions are not just simple mixing but may be complicated by biological processes from submerged macrophytes and by water joining the channel fro the hyporheic zone or from springs draining an aquifer.

Geological Inputs

The geology that underlies a river or lake has a major impact on its chemistry. A river flowing across very ancient precambrian schists is likely to have dissolved very little from the rocks and maybe similar to de-ionised water at least in the headwaters. Conversely a river flowing through chalk hills, and especially if its source is in the chalk, will have a high concentration of carbonates and bicarbonates of Calcium and possibly Magnesium.

As a river progresses along its course it may pass through a variety of geological types and it may have inputs from aquifers that do not appear on the surface anywhere in the locality.

Atmospheric Inputs

Oxygen is probably the most important chemical constituent of surface water chemistry, as all aerobic organisms require it for survival. It enters the water mostly via diffusion at the water-air interface. Oxygen's solubility in water decreases as water temperature increases. Fast, turbulent streams expose more of the water's surface area to the air and tend to have low temperatures and thus more oxygen than slow, backwaters.

Oxygen is a by-product of photosynthesis, so systems with a high abundance of aquatic algae and plants may also have high concentrations of oxygen during the day. These levels can decrease significantly during the night when primary producers switch to respiration. Oxygen can be limiting if circulation between the surface and deeper layers is poor, if the activity of animals is very high, or if there is a large amount of organic decay occurring such as following Autumn leaf-fall.

Most other atmospheric inputs come from man-made or anthropogenic sources the most significant of which are the oxides of sulphur produced by burning sulphur rich fuels such as coal and oil which give rise to acid rain. The chemistry of sulphur oxides is complex both in the atmosphere and in river systems. However the effect on the overall chemistry is simple in that it reduces the pH of the water making it more acidic. The pH change is most marked in rivers with very low concentrations of dissolved salts as these cannot buffer the effects of the acid input. Rivers downstream of major industrial conurbations are also at greatest risk. In parts of Scandinavia and West Wales and Scotland many rivers became so acidic from oxides of sulphur that most fish life was destroyed and pHs as low as pH4 were recorded during critical weather conditions.

Anthropogenic Inputs

The majority of rivers on the planet and many lakes have received or are receiving inputs from human-kind's activities. In the industrialised world, many rivers have been very seriously polluted, at least during the 19th and the first half of the 20th centuries. Although in general there has been much improvement in the developed world, there is still a great deal of river pollution apparent on the planet.

Toxicity

In most environmental situations the presence or absence of an organism is determined by a complex web of interactions only some of which will be related to measurable chemical or biological parameters. Flow rate, turbulence, inter and intra specific competition, feeding behaviour, disease, parasatism, commensalism and symbiosis are just a few of the pressures and opportunities facing any organism or population. Most chemical constituents favour some organisms and are less favourable to others. However there are some cases where a chemical constituent exerts a toxic effect. i.e where the concentration can kill or severely inhibit the normal functioning of the organism. Where a toxic effect has been demonstrated this may be noted in the sections below dealing with the individual parameters.

Chemical Constituents

Colour and Turbidity

Often it is the colour of freshwater or how clear or hazy the water is that is the most obvious visual characteristic. Unfortunately neither colour nor turbidity are strong indicators of the overall chemical composition of water. However both colour and turbidity reduce the amount of light penetrating the water and can have significant impact on algae and macrophytes. Some algae in particular are highly dependant on water with low colour and turbidity. Many rivers draining high moor-lands overlain by peat have a very deep yellow brown colour caused by dissolved humic acids.

Organic Constituents

One of the principal sources of elevated concentrations of organic chemical constituents is from treated sewage.

Dissolved organic material is most commonly measured using either the Biochemical oxygen demand (BOD) test of the Chemical oxygen demand (COD) test. Organic constituents are significant in river chemistry for the effect that they have on dissolved oxygen concentration and for the impact that individual organic species may have directly on aquatic biota. Any organic and degradable material utilises oxygen as it decomposes. Where organic concentrations are significantly elevated the effects on oxygen concentrations can be significant and as conditions get extreme the river bed may become anoxic. Some organic constituents such as synthetic hormones, pesticides, pthalates have direct metabolic effects on aquatic biota and even on humans drinking water taken from the river. Understanding such constituents and how they can be identified and quantified is becoming of increasing importance in the understanding of freshwater chemistry.

Metals

A wide range of metals may be found in rivers from natural sources where metal ores are present in the rocks over which the river flows or in the aquifers feeding water into the river. However many rivers have an increased load of metals because of industrial activities which include mining and quarrying and the processing and use of metals.

Iron

Iron, usually as Fe^{+++} is a common constituent of river waters at very low levels. As concentrations increase visible orange/brown staining appears and any increase in concentrations is likely to create conditions where complex insoluble oxides, hydroxides and carbonates of iron start precipitating out producing a semi-gelatinous and dense floc carpeting the river bed. Such conditions are very deleterious to most organisms and can cause serious damage in a river system.

Coal mining is also a very significant source of Iron both in mine-waters and from stocking yards of coal and from coal processing. Long abandoned mines can be a highly intractable source of high concentrations of Iron. Low levels of iron are common in spring waters emanating from deep-seated aquifers and maybe regarding as health giving springs. Such springs are commonly called Chalybeate springs and have given rise to a number of Spa towns in Europe and the United States.

Zinc

Zinc is normally associated with metal mining, especially Lead and Silver mining but is also a component pollutant associated with a variety of other metal mining activities and with Coal mining. Zinc is toxic at relatively low concentrations to many aquatic organisms. *Microregma* starts to show a toxic reaction at concentrations as low as 0.33 mg/l

Heavy Metals

Lead and silver in river waters are commonly found together and associated with lead mining. Impacts from very old mines can be very long-lived. In the River Ystwyth in Wales for example, the effects of silver and lead mining in the 17th and 18th centuries in the headwaters still causes unacceptably high levels of Zinc and Lead in the river water right down to its confluence with the sea. Silver is very toxic even at very low concentrations but leaves no visible of its contamination.

Lead is also highly toxic to freshwater organisms and to humans if the water is used as drinking water. As with Silver, Lead pollution is not visible to the naked eye. The River Rheidol in west Wales had a major series of lead mines in its headwaters until the end of the 19th century and its mine discharges and waste tips remain to this day. In 1919-1921 only 14 species of invertebrates were found were found in the lower Rheidol when Lead concentrations were between 0.2ppm and 0.5ppm. By 1932 the lead concentration had reduced to 0.02ppm to 0.1ppm because of the abandonment of mining and, at those concentrations, the bottom fauna had stabilized to 103 species including three leeches.

Coal mining is also a very significant source of metals, especially Iron, Zinc and Nickel particularly where the coal is rich if pyrites which oxidises on contact with the air producing a very acidic leachate which is able to dissolve metals from the coal.

Significant levels of copper are unusual in rivers and where it does it occur the source is most likely to be mining activities, coal stocking, or pig farming. Rarely elevated levels may be of geological origin. Copper is acutely toxic to many freshwater organisms, especially algae, at very low concentrations and significant concentration in river water may have have serious adverse effects on the local ecology.

Nitrogen

Nitrogenous compounds have a variety of sources including washout of oxides of nitrogen from the atmosphere, some geological inputs and some from macrophyte and algal nitrogen fixation. However for many rivers in the proximity of humans, the largest input is from sewage whether treated or untreated. The nitrogen derives from breakdown products of proteins found in urine and faeces. These products, being very soluble, often pass through sewage treatment process and are discharged into rivers as a component of sewage treatment effluent. Nitrogen may be in the form of nitrate, nitrite, ammonia or ammonium salts or what is termed albuminoid nitrogen or nitrogen still within an organic proteinoid molecule.

The differing forms of nitrogen are relatively stable in most river systems with nitrite slowly transforming into nitrate in well oxygenated rivers and ammonia transforming into nitrite/ nitrate. However, the process are slow in cool rivers and reduction in concentration may more often be attributed to simple dilution. All forms of nitrogen are taken up by macrophytes and algae and elevated levels of nitrogen are often associated with overgrowths of plants or eutrophication. These can have the effect of blocking channels and inhibiting navigation.

However, ecologically, the more significant effect is on dissolved oxygen concentrations which may become super-saturated during daylight due to plant photsynthesis but then drop to very low levels during darkness as plant respiration uses up the dissolved oxygen. Coupled with the release of oxygen in photosynthesis is the creation of bi-carbonate ions which cause a steep rise in pH and this is matched in darkness as carbon dioxide is released through respiration which substantially lowers the pH. Thus high levels of nitrogenous compounds tends to lead to eutrophication with extreme variations in parameters which in turn can substantially degrdae the ecological worth of the watercourse.

Ammonium ions also have a toxic effect, especially on fish. The toxicity of ammonia is dependent on both pH and temperature and an added complexity is the buffering effect of the blood/water interface across the gill membrane which masks any additional toxicity over about pH 8.0. The management of river chemistry to avoid ecological damage is particularly difficult in the case of ammonia as a wide range of potential scenarios of concentration, pH and temperature have to be considered and the diurnal pH fluctuation caused by photosynthesis considered. On warm summer days with high-bi-carbonate concentrations unexpectedly toxic conditions can be created.

Phosphorous

Phosphorus compounds are usually found as relatively insoluble phosphates in river water and, except in some exceptional circumstances, the origin is from agriculture of human sewage. Phosphorous can encourage excessive growths of plants and algae and contribute to eutrophication. If a river discharges into a lake or reservoir phosphate can be mobilised year after year by natural processes.

In the summer time, lakes stratify so that warm oxygen rich water floats on top of cold oxygen poor water. In the warm upper layers-the epilimnion- plants consume the available phosphate. As the plants die in the late summer they fall into the cool water layers underneath-the hypolimnion-and decompose. During winter turn-over, when a lake becomes fully mixed through the action of winds on a cooling body of water-the phosphates are spread throughout the lake again to feed a new generation of plants. This process is one of the principal causes of persistent algal blooms at some lakes.

Arsenic

Geological deposits of arsenic may be released into rivers where deep ground-waters are exploited as in parts of Pakistan. Some lead and copper mining also encounters ores containing arsenic which may then be released into local rivers.

Solids

Inert solids are produced in all montane rivers as the energy of the water helps grind away rocks into gravel, sand and finer material. Much of this settles very quickly and provides an important substrate for many aquatic organisms. Many salmonid fish require beds of gravel and sand in which to lay their eggs. Many other types of solids from agriculture, mining, quarrying, urban run-off and sewage may block-out sunlight from the river and may block interstices in gravel beds making them useless for spawning and supporting insect life.

Bacterial, Viral and Parasite Inputs

Both agriculture and sewage treatment produce inputs into rivers with very high concentrations of bateria and viruses including a wide range of pathogenic organisms. Even in areas with little human activity significant levels of bacteria and viruses can be detected originating from fish and aquatic mammals and from animals grazing near rivers such as deer. Upland waters draining areas frequented by sheep, goats or deer may also harbour a variety of opportunistic human parasites such as liver fluke. Consequently there are very few rivers from which the water is safe to drink without some form of sterilisation or disinfection. In rivers used for contact recreation such as swimming, safe levels of bacteria and viruses can be established based on risk assessment.

Under certain conditions bacteria can colonise freshwaters occasionally making large rafts of filamentous mats known as *sewage fungus*-usually *Sphaerotilus natans*. The presence of such organisms is almost always an indicator of extreme organic pollution and would be expected to be matched with low dissolved oxygen concentrations and high BOD vales.

pH

pH in rivers is affected by the geology of the water source, atmospheric inputs and a range of other chemical contaminants. pH is only likely to become an issue on very poorly buffered upland rivers where atmospheric sulphur and nitrogen oxides may very significantly depress the pH as low as pH4 or in eutrophic alkaline rivers where photosynthetic bi-carbonate ion production in photosynthesis may drive the pH up above pH10

Green Chemistry

Green chemistry, also called sustainable chemistry, is a chemical philosophy encouraging the design of products and processes that reduce or eliminate the use and generation of hazardous substances. Whereas environmental chemistry is the chemistry of the natural environment, and of pollutant chemicals in nature, green chemistry seeks to reduce and prevent pollution at its source. In 1990 the Pollution Prevention Act was passed in the United States. This act helped create a *modus operandi* for dealing with pollution in an original and innovative way. It aims to avoid problems before they happen.

As a chemical philosophy, green chemistry applies to organic chemistry, inorganic chemistry, biochemistry, analytical chemistry, and even physical chemistry. While green chemistry seems to focus on industrial applications, it does apply to any chemistry choice. Click chemistry is often cited as a style of chemical synthesis that is consistent with the goals of green chemistry. The focus is on minimizing the hazard and maximizing the efficiency of any chemical choice. It is distinct from environmental chemistry which focuses on chemical phenomena in the environment.

In 2005 Ryoji Noyori identified three key developments in green chemistry: use of supercritical carbon dioxide as green solvent, aqueous hydrogen peroxide for clean oxidations and the use of hydrogen in asymmetric synthesis. Examples of applied green chemistry are supercritical water oxidation, on water reactions and dry media reactions.

Bioengineering is also seen as a promising technique for achieving green chemistry goals. A number of important process chemicals can be synthesized in engineered organisms, such as shikimate, a Tamiflu precursor which is fermented by Roche in bacteria.

There is some debate as to whether green chemistry includes a consideration of economics, but by definition, if green chemistry is not applied, it cannot accomplish the reduction in the “use or generation of hazardous substances.”

Principles

Paul Anastas, then of the United States Environmental Protection Agency, and John C. Warner developed 12 principles of green chemistry, which help to explain what the definition means in practice. The principles cover such concepts as:

- the design of processes to maximize the amount of raw material that ends up in the product;
- the use of safe, environment-benign substances, including solvents, whenever possible;
- the design of energy efficient processes;
- the best form of waste disposal: do not create it in the first place.

The 12 principles are:

1. *Prevent waste:* Design chemical syntheses to prevent waste, leaving no waste to treat or clean up.
2. *Design safer chemicals and products:* Design chemical products to be fully effective, yet have little or no toxicity.
3. *Design less hazardous chemical syntheses:* Design syntheses to use and generate substances with little or no toxicity to humans and the environment.
4. *Use renewable feedstock:* Use raw materials and feedstock that are renewable rather than depleting. Renewable feedstock are often made from agricultural products or are the wastes of other processes; depleting feedstock are made from fossil fuels (petroleum, natural gas, or coal) or are mined.
5. *Use catalysts, not stoichiometric reagents:* Minimize waste by using catalytic reactions. Catalysts are used in small amounts and can carry out a single reaction many times. They are preferable to stoichiometric reagents, which are used in excess and work only once.
6. *Avoid chemical derivatives:* Avoid using blocking or protecting groups or any temporary modifications if possible. Derivatives use additional reagents and generate waste.
7. *Maximize atom economy:* Design syntheses so that the final product contains the maximum proportion of the starting materials. There should be few, if any, wasted atoms.

8. *Use safer solvents and reaction conditions:* Avoid using solvents, separation agents, or other auxiliary chemicals. If these chemicals are necessary, use innocuous chemicals. If a solvent is necessary, water is a good medium as well as certain eco-friendly solvents that do not contribute to smog formation or destroy the ozone.
9. *Increase energy efficiency:* Run chemical reactions at ambient temperature and pressure whenever possible.
10. *Design chemicals and products to degrade after use:* Design chemical products to break down to innocuous substances after use so that they do not accumulate in the environment.
11. *Analyze in real time to prevent pollution:* Include in-process real-time monitoring and control during syntheses to minimize or eliminate the formation of byproducts.
12. *Minimize the potential for accidents:* Design chemicals and their forms (solid, liquid, or gas) to minimize the potential for chemical accidents including explosions, fires, and releases to the environment.

Presidential Green Chemistry Challenge Awards

The Presidential Green Chemistry Challenge Awards began in 1995 as an effort to recognize individuals and businesses for innovations in green chemistry. Typically five awards are given each year, one in each of five categories: Academic, Small Business, Greener Synthetic Pathways, Greener Reaction Conditions, and Designing Greener Chemicals. Nominations are accepted the prior year, and evaluated by an independent panel of chemists convened by the American Chemical Society. Through 2006, a total of 57 technologies have been recognized for the award, and over 1000 nominations have been submitted.

In 1996, Dow Chemical won the 1996 Greener Reaction Conditions award for their 100% carbon dioxide blowing agent for polystyrene foam production. Polystyrene foam is a common material used in packing and food transportation. Seven hundred million pounds are produced each year in the United States alone. Traditionally, CFC and other ozone-depleting chemicals were used in the production process of the foam sheets, presenting a serious environmental hazard. Flammable, explosive, and, in some cases toxic hydrocarbons have also been used as CFC replacements, but they present their own problems. Dow Chemical discovered that supercritical carbon dioxide works equally as well as a blowing agent, without the need for hazardous substances, allowing the polystyrene to be more easily recycled. The CO_2 used in the process is reused from other industries, so the net carbon released from the process is zero.

In 2002, Cargill Dow (now NatureWorks) won the Greener Reaction Conditions Award for their improved polylactic acid polymerization process. Lactic acid is produced by fermenting corn and converted to lactide, the cyclic dimer ester of lactic acid using an efficient, tin-catalyzed cyclization. The L,L-lactide enantiomer is isolated by distillation and polymerized in the melt to make a crystallizable polymer, which has use in many applications including textiles and apparel, cutlery, and food packaging. Wal-Mart has announced that it is using/will use PLA for its produce packaging. The NatureWorks PLA process substitutes renewable materials for petroleum feedstocks, doesn't require the use of hazardous organic solvents typical in other PLA processes, and results in a high-quality polymer that is recyclable and compostable.

In 2003 Shaw Industries was recognized with the Designing Greener Chemicals Award for developing EcoWorx Carpet Tile. Historically, carpet tile backings have been manufactured using bitumen, polyvinyl chloride (PVC), or polyurethane (PU). While these backing systems have performed satisfactorily, there are several inherently negative attributes due to their feedstocks or their ability to be recycled. Shaw selected a combination of polyolefin resins as the base polymer of choice for EcoWorx due to the low toxicity of its feedstocks, superior adhesion properties, dimensional stability, and its ability to be recycled.

The EcoWorx compound also had to be designed to be compatible with nylon carpet fiber. Although EcoWorx may be recovered from any fiber type, nylon-6 provides a significant advantage. Polyolefins are compatible with known nylon-6 depolymerization methods. PVC interferes with those processes. Nylon-6 chemistry is well-known and not addressed in first-generation production. From its inception, EcoWorx met all of the design criteria necessary to satisfy the needs of the marketplace from a performance, health, and environmental standpoint. Research indicated that separation of the fiber and backing through elutriation, grinding, and air separation proved to be the best way to recover the face and backing components, but an infrastructure for returning postconsumer EcoWorx to the elutriation process was necessary. Research also indicated that the postconsumer carpet tile had a positive economic value at the end of its useful life. EcoWorx is recognized by MBDC as a certified Cradle to Cradle design.

In 2005, Archer Daniels Midland (ADM) and Novozymes N.A. won the Greener Synthetic Pathways Award for their enzyme interesterification process. In response to the U.S. Food and Drug Administration (FDA) mandated labeling of *trans*-fats on nutritional information by January 1, 2006, Novozymes and ADM worked together to develop a clean, enzymatic process for the interesterification of oils and fats by interchanging saturated and unsaturated fatty acids. The result is commercially viable products without *trans*-fats. In addition to the human health benefits of eliminating *trans*-fats, the process has reduced the use of toxic chemicals and water, prevents vast amounts of byproducts, and reduces the amount of fats and oils wasted.

In 2006, Professor Galen J. Suppes, from the University of Missouri–Columbia, was awarded the Academic Award for his system of converting waste glycerin from biodiesel production to propylene glycol. Through the use of a copper-chromite catalyst, Professor Suppes was able to lower the required temperature of conversion while raising the efficiency of the distillation reaction. Propylene glycol produced in this way will be cheap enough to replace the more toxic ethylene glycol that is the primary ingredient in automobile antifreeze.

Other Awards

The Royal Australian Chemical Institute (RACI) presents Australia's Green Chemistry Challenge Awards. This awards program is similar to that at the United States Environmental Protection Agencyý (EPA), although the Institute has included a category for Green Chemistry education as well as Small Business and Academic or Government. The Canadian Green Chemistry Medal is an annual award given to an individual or group for promotion and development of green chemistry in Canada and internationally. The winner is presented with a citation recognizing the achievements together with a sculpture.

Green Chemistry activities in Italy centre around an inter-university consortium known as INCA. Beginning in 1999, the INCA has given three awards annually to industry for applications of green chemistry. The winners receive a plaque at the annual INCA meeting.

In Japan, The Green & Sustainable Chemistry Network(GSCN), formed in 1999, is an organization consisting of representatives from chemical manufacturers and researchers. In 2001, the organization began an awards program. GSC Awards are to be granted to individuals, groups or companies who greatly contributed to green chemistry through their research, development and their industrialization. The achievements are awarded by Ministers of related government agencies.

In the United Kingdom, the Crystal Faraday Partnership, a non-profit group founded in 2001, awards businesses annually for incorporation of green chemistry. The Green Chemical Technology Awards have been given by Crystal Faraday since 2004; the awards were presented by the Royal Society of Chemistry prior to that time. The award is given only to a single researcher or business, while other notable entries are given recognition as well.

The Nobel Prize Committee recognized the importance of green chemistry in 2005 by awarding Yves Chauvin, Robert H. Grubbs, and Richard R. Schrock the Nobel Prize for Chemistry for "the development of the metathesis method in organic synthesis." The Nobel Prize Committee states, "this represents a great step forward for 'green chemistry', reducing potentially hazardous waste through smarter production. Metathesis is an example of how important basic science has been applied for the benefit of man, society and the environment."

Trends

Attempts are being made not only to quantify the *greenness* of a chemical process but also to factor in other variables such as chemical yield, the price of reaction components, safety in handling chemicals, hardware demands, energy profile and ease of product workup and purification. In one quantitative study, the reduction of nitrobenzene to aniline receives 64 points out of 100 marking it as an acceptable synthesis overall whereas a synthesis of an amide using HMDS is only described as adequate with a combined 32 points. Green chemistry is increasingly seen as a powerful tool that researchers must use to evaluate the environmental impact of nanotechnology. As nanomaterials are developed, the environmental and human health impacts of both the products themselves and the processes to make them must be considered to ensure their long-term economic viability.

Laws

In 2007, Europe put into place the Registration, Evaluation, Authorisation, and Restriction of Chemicals (REACH) program, which requires companies to provide data showing that their products are safe. The US Toxic Substances Control Act, passed in 1976, has no similar provisions. On September 29, 2008 California approved two laws which encourage green chemistry, launching the California Green Chemistry Initiative. The law requires California's Department of Toxic Substances Control to prioritize "chemicals of concern", and puts the burden of testing on the agency rather than industry. The laws were criticized by Paul Anastas, who stated that the laws did not go far enough in encouraging research, education, and industry incentives.

Examples

Supramolecular chemistry: Research is currently ongoing in the area of supramolecular chemistry to develop reactions which can proceed in the solid state without the use of solvents. The cycloaddition of *trans*-1,2-bis(4-pyridyl)ethylene is directed by resorcinol in the solid state. This solid-state reaction proceeds in the presence of UV light in 100% yield.

Reducing market barriers: In March 2006, the University of California published a report by Dr. Michael P. Wilson and colleagues, Daniel A. Chia and Bryan C. Ehlers, on potential chemicals policy for the California Legislature entitled, Green Chemistry in California: A Framework for Leadership in Chemicals Policy and Innovation. The report finds that long-standing weaknesses in the U.S. chemical management program, notably the Toxic Substances Control Act (TSCA) of 1976, have produced a chemicals market in the U.S. that discounts the hazardous properties of chemicals relative to their function, price, and performance. The report concludes that these market conditions represent a key barrier to the scientific, technical, and commercial success of green chemistry in the U.S., and that fundamental policy changes are needed to correct these weaknesses.

Education: In 2005, the University of Oregon's chemistry department unveiled the Greener Education Materials, a database of green chemistry topics.

Laboratory chemicals: Several laboratory chemicals are controversial, and alternatives have been explored. The Massachusetts Institute of Technology has created the Green Alternatives Wizard to help identify alternatives. Ethidium bromide, xylene, mercury, and formalin have been identified as "worst offenders" which have alternatives.

Chapter 6

Photochemistry

Photochemistry is concerned with the absorption, excitation and emission of photons by atoms, atomic ions, molecules, molecular ions, etc. The simplest photochemical process is seen with the absorption and subsequent emission of a photon by a gas phase atom such as sodium. When the sodium atom absorbs a photon it is said to be excited. After a short period of time, the excited state sodium atom emits a photon of 589 nm light and falls back to the ground state:

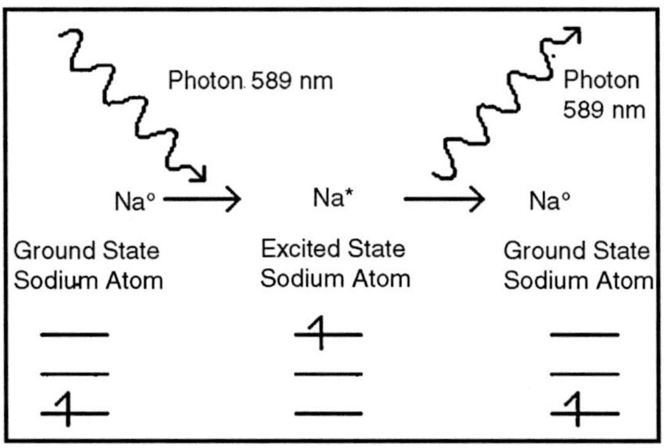

The atom can be excited by a flame, and this is the basis of the well known "flame test" of Group 1 and 2 metal salts (LiCl, NaBr, CaCl2, SrCl2, etc.) and atomic emission spectroscopy.:

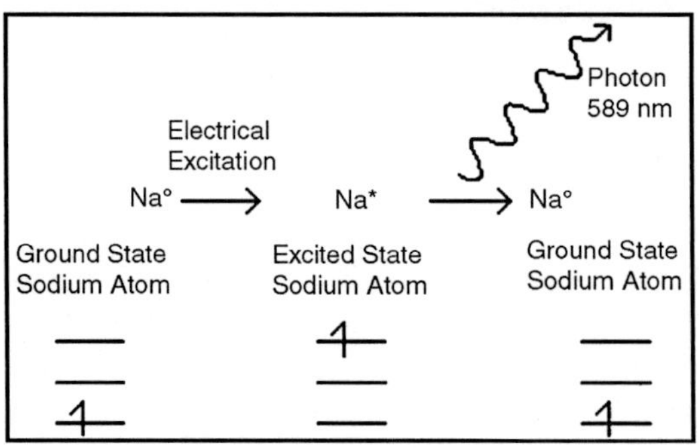

Concept

The study of chemical reactions, summarizations and physical behaviour that may occur under the influence of visible and/or ultraviolet light is called Photochemistry.

Two fundamental principles are the foundation for understanding photochemical transformations:

1. The first law of photochemistry, the Grotthuss-Draper law, states that light must be absorbed by a compound in order for a photochemical reaction to take place.
2. The second law of photochemistry, the Stark-Einstein law, states that for each photon of light absorbed by a chemical system, only one molecule is activated for subsequent reaction. This "photoequivalence law" was derived by Albert Einstein during his development of the quantum (photon) theory of light.

The efficiency with which a given photochemical process occurs is given by its Quantum Yield (Φ). Since many photochemical reactions are complex, and may compete with unproductive energy loss, the quantum yield is usually specified for a particular event. Thus, we may define quantum yield as "the number of moles of a stated reactant disappearing, or the number of moles of a stated product produced, per einstein of monochromatic light absorbed.", where an Einstein is one mole of photons. For example, irradiation of acetone with 313 nm light (3130 Å) gives a complex mixture of products, as shown in the following diagram. The quantum yield of these products is less than 0.2, indicating there are radioactive (fluorescence and phosphorescence) and non-radiative return pathways (green arrow).

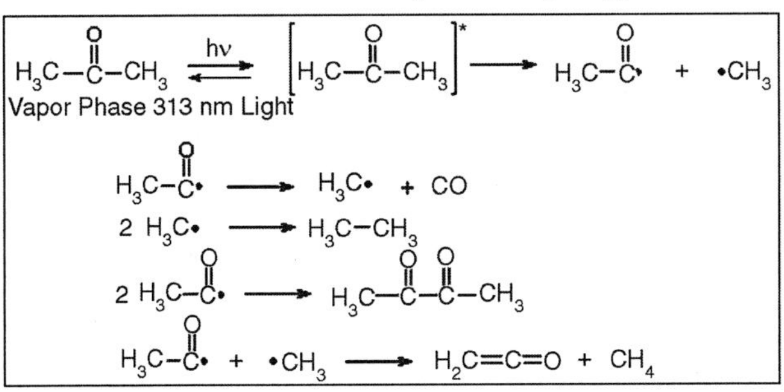

Several secondary radical reactions then follow (shown in the gray box), making it difficult to assign a quantum yield to the primary reaction. The biacetyl product, formed in the third reaction, may itself be excited by light or accept excitation energy from an excited acetone molecule, further complicating this process. By comparison, the light induced chlorination of methane, or other alkanes, has a large quantum yield, often near 10^6, because of the secondary chain reactions that follow the primary cleavage of the Cl-Cl bond. Equations illustrating such halogenation reactions have been presented elsewhere in this text.

Absorption of visible and/or ultraviolet light by a molecule introduces energy sufficient to break or reorganize most covalent bonds. From the relationship $E = hc/\lambda$, we see that longer wavelength visible light (400 to 800 nm) is less energetic (70 to 40 kcal/mole) than light in the accessible shorter wavelength (200 to 400 nm) near ultraviolet region (150 to 70 kcal/mole). Consequently, ultraviolet light is most often used to effect photochemical change. Care must also be taken to construct lamps and reaction vessels from glass that is transparent to the desired wavelength range. The low wavelength cut-off for some common glass types are given in the table on the right.

The light required for a photochemical reaction may come from many sources. Giacomo Ciamician, regarded as the "father of organic photochemistry", used sunlight for much of his research at the University of Bologna in the early 1900's. Depending on the compounds being studied and the information being sought, bright incandescent lamps (chiefly infrared and visible light), low, medium and high pressure mercury lamps (185 – 255 nm, 255 –1000 nm & 220 –1400 nm respectively), high intensity flash sources and lasers have all been used. In careful studies of specific chromophores, sources of monochromatic light may be desired.

As a rule, such excitation results in a change in molecular orbital occupancy, an increase in energy, and changes in local bonding and charge distribution. The principles set forth in the UV-Visible Spectroscopy will provide a helpful foundation.

Type of Glass	Wavelength Cut-off
Pyrex	< 275 nm
Corex	< 260 nm
Vycor	< 220 nm
Quartz	< 170 nm

We begin by considering the electronic excitation of a simple diatomic molecule such as Cl_2 or Br_2. Both absorb light, chlorine in the 300 to 380 nm region. and bromine in the 360 to 510 nm region. Both the ground (lowest energy electronic state) and excited states are shown as energy profiles populated by vibrational energy states (green lines) as well as rotational states (not shown). The electron reorganization that occurs when the ground electronic state is excited by absorption of a photon takes place much more rapidly than any movement of the atom nuclei that eventually follow.

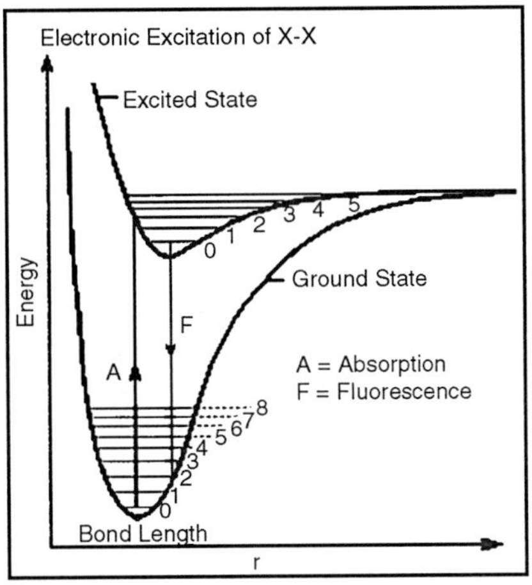

In other words, electron shifts, when viewed from the perspective of the nuclear coordinates, occur as if the heavier nuclei were fixed in place. This consequence of the Born-Oppenheimer approximation led James Franck and R. Condon to formulate the Franck-Condon Principle: Electronic transitions occur much faster than nuclei can respond. Overall bonding in an excited state is usually lower than in the ground state. Thus, the X–X bond length is increased in the excited state. At normal temperatures essentially all molecules will exist in the ground vibrational state (zero level).

Virtually all organic compounds have more than two atoms, so the potential energy state diagram of X_2 must be adjusted for the increased number of bonding relationships. One way of doing this is to retain the energy coordinate while dispensing with the dimensional (r) coordinate.

The Jablonski diagram shown below is an example, in which the spatial orientation of the various electronic states is not specified. Nevertheless, Franck-Condon transitions are expected. One important feature conveyed by the diagram is that more than one electronic excited state is likely to exist for a given molecule, six are drawn and labeled in the diagram. Each electronic state will have a group of vibrational (and rotational) states, depicted by light blue lines above each state marker.

Excited states may be classified as singlet or triplet based upon their electron spin angular momentum. The electrons in most non-metallic organic compounds are paired (opposite spins) in bonding and non-bonding orbitals, resulting in a net zero spin diamagnetic molecule for the ground state. Such states have a single energy state in an applied magnetic field, and are called singlets. Electronic states in which two electrons with identical spin occupy different orbitals (the Pauli exclusion principle) have a net spin of 1 (2 × 1/2) and are paramagnetic. In a magnetic field such states have three energy levels (+1, 0, −1) and are called triplets.

Molecular oxygen is a rare example of a triplet ground electronic state. The distinction between singlet and triplet states is important because photon induced excitation always leads to a state of the same multiplicit, *i.e.* singlet to singlet or triplet to triplet. Since most ground states are singlets, this means that the excited states initially formed by absorption of light must also be singlets. Internal conversion of excited states to lower energy states of the same multiplicity takes place rapidly with loss of heat energy (relaxation). Alternatively, an excited state may return to the ground state by emitting a photon (radiative decay). In the study of acetone described above, nearly 80% of the excited singlet states lose energy by internal conversion and about 3% by fluorescence. Conversion of a singlet state to a lower energy triplet state, or vice versa, is termed intersystem crossing and is slower than internal conversion. Radiative decay from a triplet state is called phosporescence and is generally quite slow.

Laws of Photochemistry

Basic

Photochemistry is the study of light-induced chemical reactions and physical processes. A photochemical event involves the absorption of light to create an excited species that may subsequently undergo a number of different reactions. These include unimolecular reactions such as dissociation, ionization, and isomerization; bimolecular reactions, which involve a reaction with a second molecule or atom to form a new compound; and reactions producing an emission of light, or luminescence. A photochemical reaction differs notably from a thermally, or heat, induced reaction in that the rate of a photochemical reaction is frequently greatly accelerated, and the products of the photochemical reaction may be impossible to produce otherwise. With the advent of lasers(powerful, single-colour light sources) the field of photochemistry has advanced tremendously over the past few decades.

An increased understanding of photochemistry has great implications outside of the laboratory, as photochemical reactions are an extremely important aspect of everyday life, underlying the processes of vision, photosynthesis, photography, atmospheric chemistry, the production of smog, and the destruction of the ozone layer. The absorption of light by atoms and molecules to create an excited species is studied in the field of spectroscopy. The study of the reactions of this excited species is the domain of photochemistry. However, the fields are closely related; spectroscopy is routinely used by photochemists as a tool for identifying reaction pathways and products and, recently, for following reactions as they occur in real time. Some lasers can produce a pulse of light that is only "on" for 1 femtosecond (10^{-15}seconds). A femtosecond laser can be used like an extremely high-speed strobe camera to spectroscopic ally "photograph" a photochemical reaction.

The separation of an electron from an atom or molecule, leaving a positively charged ion, is a special form of dissociation called ionization. Ionization following absorption of a photon (path ii) usually occurs with light of very short wavelengths (less than 100 nm) and therefore is usually not studied by photochemists, although it is of great importance in x-ray technology. X rays are also sometimes referred to asionizing radiation.

First Law

Absorption of light is just the first step in a chain of events leading to a photochemical change. However, it is an essential step. Light that is transmitted, unchanged, through a medium can have no effect on the medium through which it passed. This principle, sometimes referred to as the First Law of Photochemistry was first advanced nearly two centuries ago by Grothus and Draper in 1818. Only absorbed photons can produce photochemical changes. While this may seem obvious, it leads to some interesting predictions. For example, because Beer's Law tells us that absorption is proportional to the incident light intensiy, as long as path length and absorber concentration don't change, then the First Law predicts that a measured reaction product should be linearly related to the number of incident photons. This means that if you double the incident light intensity, you should be able to observe the same photochemical effect with only half the duration of illumination, a phenomenon sometimes referred to as "reciprocity." Reciprocity can be observed in many photochemical and photobiological processes, but must be experimentally verified.

The Jablonski Diagram

The energy gained by a molecule when it absorbs a photon causes an electron to be promoted to a higher electronic energy level. The transitions among molecular excited states can be described by the energy level diagram that was introduced in the Basic Photophysics module. The symbols S_0, S_1, T_2 etc. refer to the ground electronic state (S_0), first excited singlet state (S_1), second excited triplet state (T_2) and so on. The horizontal lines represent vibrational levels of each electronic state. Straight arrows indicate radiative transitions and, curly arrows, non-radiative transitions. The boxes detail the electronic spins in each orbital with electrons shown as up and down arrows to distinguish their spin. In this all transitions from one electronic state to another originate from the lowest vibrational level of the initial electronic state. For example, fluorescence occurs only from S_1 because the higher singlet states (S_2 etc) decay so rapidly by internal conversion that fluorescence from these states cannot compete.

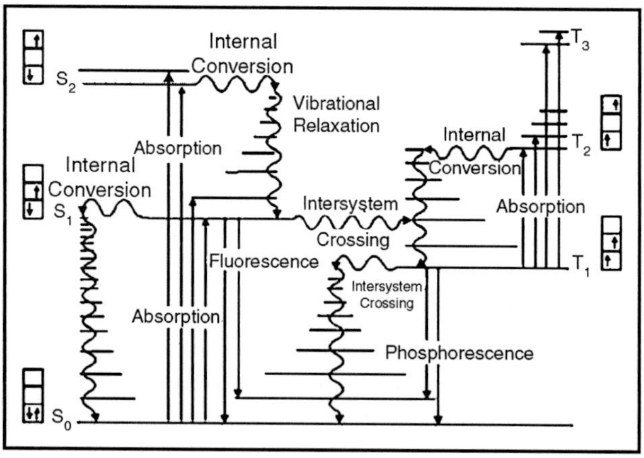

Orbital Energy Level Diagrams

The physical and chemical properties of a molecule are determined primarily by the distribution of electrons associated with the molecule. Reactions that take place photochemically (but not thermally) result from the new electronic distribution of the excited state produced by photon absorption. For organic chromophores, the absorption of light usually results in the excitation of electrons in π bonding molecular orbitals or non-bonding atomic orbitals (denoted n) to antibonding π^* molecular orbitals.

The less energetic σ molecular orbitals are not usually involved. (Antibonding orbitals are those in which the forces holding a molecule together are weaker than in the corresponding bonding orbital. That is, the molecule is less stable. See Suggested Readings at the end of this module for sources of additional information on bonding and antibonding molecular orbitals.)

Quantum Yields

Formula

The quantum yield (ϕ) of a process in which molecules give up their excitation energy (or "decay") is the fraction of excited molecules that decay via that pathway (Clayton 1971, 1980). Mathematically, the quantum yield of a process, such as photochemistry, is defined as follows:

$$\Phi = \frac{\text{yield od photochemical products}}{\text{tota l number of quanta absorbed}}$$

The value of ϕ for a particular process can range from 0 (if that process is never involved in the decay of the excited state) to 1.0 (if that process always deactivates the excited state). The sum of the quantum yields of all possible processes is 1.0. In functional chloroplasts kept in dim light, the quantum yield of photochemistry is approximately 0.95, the quantum yield of fluorescence is 0.05 or lower, and the quantum yields of other processes are negligible. The vast majority of excited chlorophyll molecules therefore lead to photochemistry. The quantum yield of formation of the products of photosynthesis, such as O_2, can be measured quite accurately. In this case the quantum yield is substantially lower than the value for photochemistry, because several photochemical events must take place before any O_2 molecules form.

For O_2 production the measured maximum quantum yield is approximately 0.1, meaning that 10 quanta are absorbed for each O_2 molecule released. The reciprocal of the quantum yield is called the quantum requirement. The minimum quantum requirement for O_2 evolution is therefore about 10. Quantitative measurements of the absorption of light and the fate of the energy contained in the light are essential to an understanding of photosynthesis.

Basic

A molecule in one of several ways subsequently loses the energy that a molecule gains when it absorbs light. As shown graphically in the Jablonski diagram, above, it can lose the energy as heat as it returns to the ground state (internal conversion). Alternately, it can lose the energy as light (fluorescence).

A third pathway is intersystem crossing to a triplet state from which energy can be lost as light (phosphorescence). And finally, the energy can be transferred to another molecule. (Energy transfer to oxygen will be covered in more detail in the Basic Photosensitization module.) Of course, for a single photon absorbed by a single molecule, only one pathway is followed.

But, if we study many such instances we can define the probability of occurrence for each. This is what we call the quantum yield of a process. It is the probability that an absorbed photon undergoes a particular process. Thus, we can define a quantum yield for fluorescence, a quantum yield for phosphorescence, or a quantum yield for other pathways.

Each quantum yield is typically a number between 0 and 1 (except under unusual circumstances) and the total of all quantum yields for a particular absorption process should sum to one. Note that these processes are competing. If conditions are altered such that the quantum yield for fluorescence is increased, for example, then the quantum yield for some other process (es) must decrease. We can formalize the discussion above by considering a molecule, M, that is exposed to light and absorbs photons at the rate I_{abs}.

As shown by the succeeding lines of the following relationships, the excited singlet state of the molecule, $^1M^*$ can fluoresce emitting a photon, $h\nu_f$. It can lose energy as heat and move to the triplet excited state $^3M^*$ by intersystem crossing. Finally, it can lose energy as heat and move to the ground state by internal conversion.

The rate of each of these processes will be proportional to the concentration of the excited singlet state, $^1M^*$, and a rate constant for each process, as given in the 2nd column below.

	Rate
$M \rightarrow {}^1M^*$	I_{abs}
${}^1M^* \xrightarrow{k_f} M + h\nu_f$	$k_f[{}^1M^*]$
${}^1M^* \xrightarrow{k_{isc}} {}^3M^* + \text{heat}$	$k_{isc}[{}^1M^*]$
${}^1M^* \xrightarrow{k_{ic}} M^* + \text{heat}$	$k_{ic}[{}^1M^*]$

Because the lifetime of the singlet excited state is relatively short, we can assume that all of the excited singlet states that are formed by light absorption will rapidly decay through one of the three means just described.

This is the "steady state" approximation. Since the rate of formation of the excited singlet state is I_{abs}, then the sum of the rates of loss, each given above, must be equal to the rate of formation.

$$I_{abc} = \sum_i k_i[^1M^*]$$

From this relationship we can determine the quantum yield for each process. It is simply the rate of that process as a fraction of all pathways for loss of energy.

Since we know that the rate for all pathways of energy loss is equal to the rate of energy gain by absorption, as we've just discussed, then we can determine quantum yield as follows:

$$\Phi = \frac{\text{rate of process}}{I_{abc}}$$

$$\therefore \Phi_f = \frac{K_f[^1M^*]}{I_{abc}} = \frac{K_f[^1M^*]}{\sum_i k_i[^1M^*]} = \frac{k_f}{\sum_i k_i} = k_f\tau_f$$

$$\tau_f = \text{obserced Flourescence (sin glet) lifetime}$$

$$= \frac{1}{\sum_i k_i}$$

$$\therefore \Phi_{isc} = \frac{k_{isc}[^1M^*]}{I_{abc}} = \frac{k_{isc}[^1M^*]}{\sum_i k_i[^1M^*]} = \frac{k_{isc}}{\sum_i k_i} = k_{isc}\,\tau_f$$

etc...

$$\tau_f^0 = \text{raditive lifetime}$$

$$= \frac{1}{k_f}$$

In the derivations shown above, we define both the fluorescence lifetime and the radiative lifetime. It may, at first, seem paradoxical that the fluorescence lifetime is not the inverse of the fluorescence rate constant, but is the inverse of the sum of all rate constants. However, a moment's reflection can clarify the situation. What determines how long fluorescence will continue is how long the excited singlet state exists. The longer it exists, the longer fluorescence will persist. Thus, the fluorescence lifetime is dependent on all processes that deplete the excited singlet state. If there are no other competing processes, that is if all energy is lost only by fluorescence, then the expression for the observed fluorescence lifetime decays to $1/k_f$, the radiative lifetime, τ_f^o.

Primary and Secondary Photochemical Process

Photochemical Processes

Ionization $A^* \rightarrow A+ + e-$

Electron transfer $A^* + B \rightarrow A+ + B-$ or $A- + B+$

Dissociation $A^* \rightarrow B + C$ or $A^* + B\text{-}C \rightarrow A + B + C$

Addition $2A^* \rightarrow B$

Abstraction $A^* + B\text{- }C \rightarrow A\text{- }B + C$

Isomerization or rearrangement $A^* \rightarrow A'$

Is thus constant and the reaction rate must be independent of the concentration of the light absorbing component. If, on the other hand, the reaction is studied in an infinitely thin layer or a and c (the concentration) are small, then the above expression is reduced to

/b =/<> acx

and, accordingly, the reaction must be unimolecular with respect to the light absorbing component. These two extreme cases were found to be in good agreement with experimental results available at that time, but the known dependence of several reactions on the concentration of the non-absorbing components required additional assumptions. The whole treatment was therefore somewhat formal, since the main kinetic question: why on illumination only certain molecules of the light absorbing component reacted and not its whole mass at once, could not be answered satisfactorily.

Light Quanta* A starting point for the rapid development of photochemistry along new lines was undoubtedly the introduction of quantum theory. General attention turned from rather unsuccessful therinodymunic speculations to a quantitative study of photochemical kinetics, resulting in a deeper understanding of the role of radiant energy in producing chemical changes and in many valuable contributions to the general theory of chemical kinetics.

An analysis of the modern literature on the subject shows that by far the larger proportion of all the theoretical and experimental photochemical studies in the course of the last fifteen years have been connected to some extent with Einstein's equivalence law and though it soon became apparent that this is by no means an exact law governing all photochemical reactions, nevertheless, the idea of light absorption according to quantum laws, applied to photochemistry, proved to be at least a most useful working hypothesis. Moreover, the study of photochemical reactions gained, by application of the equivalence law, subsequently discussed in detail, an important advantage over that of thermal reactions.

Whereas these latter meet difficulties in solving two main kinetic problems: how many and from win** ourees of energy activated molecules are formed in the course of a reaction and how do they react, phrtochemistry has a ready answer to the first of these questions, in so iar as the assumptions of the equivalence law be accepted.

Accordingly, an answer to the second question can be sought and found in a more satisfying manner and, actually, up to the present independently of the change in the electronic system. Franck further arrives at the conclusion that, if an adiabatic dissociation of hetero polar molecules does occur, then ions must be formed and not atoms, since the ions and not the atoms are the oscillating parts of the molecule.

The equivalence law is strictly valid only for ideal processes and the deviations of the real ones are due not so much to the secondary reactions (of Bodenstein's type) as to the fact, normally ignored, that the whole Bohr concept of light absorption is strictly valid only for atoms and molecules at infinite dilution. In real photochemical reactions we deal usually with concentrated systems and here the mutual influences of neighboring molecules play an unknown but important role. The deviations from Beer's absorption law, frequently observed, are only an indication of this effect.

Laws of Photochemical

As has been already pointed out, the idea that light is absorbed by photochemically active molecules in definite energy quanta seems to be now almost generally accepted as the basis for the interpretation of the kinetics of photochemical reaction. Opinions on the nature of the primary and secondary processes arc, however, widely varying and, though many of the theories described found a successful application in the treatment of the experimental material, it would be an im- possible task to point out at present a theory that will account for all the experimental data. Among the large variety of recently investigated and reinvest!- gated photochemical reactions, a certain rather limited number ap- pears, from the point of view of quantum absorption, to offer less difficulties in the quantitative explanation of reaction kinetics. These reactions are usually classified as reactions following the equivalence law and they will be the chief object of consideration in the present section.

Assumptions made in the derivation of the equivalence law require that the rate of a photochemical reaction shall be independent of the concentration of the molecules present and of the temperature of the reacting body; the rate must be fully determined by the amount of absorbed light energy. Furthermore, one gram mol of the light absorbing component must react on absorption of fm amount of light energy equal to Nhv* Thus, the rate of reaction per unit (calorie) of absorbed energy must depend on the wave length of active radiation.

We find l that A × 3,512 × 10~ gram mols of a substance must react according to the equivalence law on absorption of one calorie of light energy of the wave length x. This relation shows that, with shorter wave length, the efficiency per gram calorie of radiation energy should decrease. Warburg's Experimental Researches, among the first who under- took the experimental verification of the equivalence law was War-burg, His extensive researches include a large variety of photochemical reactions, both in the gaseous phase and in solutions?. His work undoubtedly must be considered as classical with regard to experimental skill and thorough theoretical treatment of the subject.

The results obtained by Warburg were in general not very favorable to a strict application of the equivalence law, but two of the reactions studied proved to be in relatively close agreement with the theoretical predictions of Einstein, These are the decompositions of gaseous hydrogen iodide and bromide.

Hydrogen-Iodide Decomposition, This reaction was discovered in 1877 by Lemoine and later studied by Bodonstein, 3 who used sunlight as the source of illumination and found a unimolecular reaction. The absorption spectrum of hydrogen iodide and its decomposition under the action of ultra-violet light were then studied by Coehn and Stuckard. 3 The absorption starts, according to these authors, at about 3300A. arid increases for shorter wave lengths. Under the action of light of wave lengths longer than 2500A. is complete and only light of shorter wave length causes a partial formation of hydrogen iodide at about 270 C., the temperature of their experiments.

Warburg 4 studied the reaction at room temperature and used monochromatic light of three different wave lengths as the source of illumination. The amount of absorbed light energy varied in these experiments in the ratio 1:2.2 and the partial pressure of hydrogen iodide, which was mixed with hydrogen, varied from 80 to 244 mm. The rate of reaction was found to be dependent only on the amount of absorbed light energy. Recently Bodenstein and Lieneweg a extended the experiments of Warburg and measured the rate of reaction under widely varying conditions. They found that an increase in temperature from 150 to 175 left the rate of reaction unchanged.

Nitrogen also was without influence even at a pressure of 3.5 atmospheres. In presence of iodine, the rate of reaction is not accelerated, but retarded, this effect being due to the absorption of active radiation by iodine.

Even liquid hydrogen iodide decomposes at room temperature at the rate of two molecules per absorbed light quantum. Discussing the possible explanation of the observed quantum yield of two molecules, Warburg suggested that hydrogen iodide is decom- posed into atoms on absorption of quantum energy of the active radiation.

Warburg further suggested that the atoms formed could react in one of the following ways:

- II + HI –MI. + I
- I + HI–*– I a + H
- H + H -> It,
- 11 + I –*HI
- I + I –M,

Primary and Secondary Process

A primary photochemical process is one in which an excited molecule gives rise directly to the final reaction products (*e.g.*, *cis-trans* photoisomerization of retinal).

A secondary photochemical process is one in which an excited molecule gives rise to an intermediate, which then takes part in "dark" reactions that lead to the final reaction products. There are many photophysical processes which may deactivate excited molecules. Whether or not a photochemical process will occur depends on the timescales associated with formation and deactivation of the excited state and how they relate to reaction rates.

Kinetics of Photochemical Reactions

The photoreaction of iodine in solution has long been the model system of choice for the study of reaction kinetics. In particular, I_2 in CCl_4 is very important due to its reduced-rate kinetics.

A method has been developed to resolve the structural dynamics of a photo excited system by pulsed X-ray scattering, that allows a natural time resolution of 100 ps. Stroboscopic data acquisition is achieved by synchronizing a femtosecond laser to the chopped X-ray pulses from an in-vacuum undulator at ID09 using its full emission spectrum at 3% bandwidth. The high flux of 10^{11} photons per second at 900 Hz pulse rate is indispensable regarding the low photo induced modulation signals of the full scattering in the intensity range of 10^{-4}.

The information content differs depending on the region in q space considered. At high momentum transfer, where strongly correlated atomic structures manifest, the distance changes of the iodine atoms can be observed. A mixture of states is observed within the first 2 nanoseconds, the stretched molecule (so-called A/A' state) and iodine atoms from the dissociation reaction. Occupancies and lifetimes can be inferred from the time-resolved q scans. The collapse of the cage around a single iodine atom causes the solvent to come closer to an iodine scattering centre, whereas the extended I_2 state pushes the solvent outwards. The sum of all contributions is seen, which represents the Fourier inversion of the difference scattering data. The interpretation is confirmed by molecular dynamics (MD) studies red line as seen. For this purpose a MD simulation has been performed with a box of 256 CCl_4 molecules and one iodine molecule using appropriate pair potentials and steps of 1 fs in a 1ns observation interval.

At larger time delays hydrodynamics comes into play as the energy flow from the excited iodine centres to the solvent is completed and the pressure in the excited region relaxes. At about one microsecond the system is equilibrated, both thermally and mechanically, with recombined iodine molecules and the heated solvent, whose difference scattering signal persists. Thus the solvent acts as a calorimeter that mirrors the energy deposited into the system by the photoreaction. This energy can be compared to the number of excited iodine molecules to give information about ultrafast geminate reactions beyond the time resolution. The challenge for the future will be to extend the procedure to larger molecules and biologically relevant aqueous solutions of photoactive molecules.

Photosensitive Reaction

Abnormal responses to sunlight or artificial light due to extreme reactivity of light-absorbing molecules in tissues. It refers almost exclusively to skin photosensitivity, including sunburn, reactions due to repeated prolonged exposure in the absence of photosensitizing factors, and reactions requiring photosensitizing factors such as photosensitizing agents and certain diseases. With restricted reference to skin tissue, it does not include photosensitivity of the eye to light, as in photophobia or photosensitive epilepsy.

Fluorescence and Phosphorescence

Fluorescence and phosphorescence are both examples of photolumin-escence. In both cases, light is absorbed and then re-emitted at a less-energetic wavelength. Fluorescent and phosphorescent objects usually glow under exposure to ultraviolet light (though this is not a requirement). Fluorescence involves absorbing and releasing lower energy light almost immediately, while the light release of phosphorescence is delayed, so these materials appear to glow in the dark.

Fluorescence

Fluorescence is a phenomenon that causes a substance to "glow" in the visible spectrum when exposed to ultraviolet light. Minerals that exhibit fluorescence are known as "fluorescent minerals". Fluorescent minerals contain particles in their structure known as *activators*, which respond to ultraviolet light by giving off a visible glow. Ultraviolet light is a form of electromagnetic radiation invisible to the human eye. It is given off by the sun and by common fluorescent lamps used for lighting, but they also give off considerable white light (visible light), preventing the fluorescence from being seen. The ultraviolet reaction is visible with a special fluorescent lamp with a filter that blocks white light but allows ultraviolet light to pass through. This lamp is known as an *ultraviolet fluorescent lamp*, or *UV lamp*. Obviously, the reaction will only be visible in a dark area, where the presence of white light is weak.

There are two classified ultraviolet wavelengths: Longwave and shortwave.Some minerals fluoresce the same colour in both wavelengths, others fluoresce in only one wavelength, and yet others fluoresce different colours in different wavelengths.Some UV lamps have two separate filters: one for longwave and the other for shortwave. Many more minerals fluoresce in shortwave than in longwave; only a fairly small amount of minerals fluoresce in longwave.Longwave fluorescent lamps are fairly inexpensive, whereas shortwave lamps can be costly. When this guide labels a mineral as fluorescent without specifying longwave or shortwave, shortwave fluorescence is implied.

Fluorescence is not a reliable method for mineral identification, since certain minerals of the same species may fluoresce different colours from different localities. Once a mineral has been identified, its fluorescent colour can sometimes be used to identify its place of origin. Similarly, if an unidentified mineral from a certain locality glows a specific colour, it can often be identified by the colour of its fluorescence.

Colour and intensity of the fluorescence varies among specimens of a particular mineral. However, specimens from the same locality virtually always fluoresce the same colour. When a fluorescent lamp is lit, never look at the light source, as it can damage the eyes permanently. In addition, skin should not be exposed to the light source for extended periods, as it can cause sunburns and long term skin problems.

Fluorescence is the process whereby a molecule in the lower of two electronic states (generally the ground state) is excited to a higher electronic state by radiation whose energy corresponds to an allowed absorption transition, followed by the emission of radiation as the system decays back to the original state. The decay process can follow several pathways. If the decay is back to the original lower state, the process is called resonance fluorescence and occurs rapidly, in about one nanosecond.

Resonance fluorescence is generally observed formonatomic gases and for many organic molecules, in particular aromatic systems that absorb in the visible and near-ultraviolet regions. For many molecules, especially aromatic compounds whose electronic absorption spectra lie predominately in the shorter-wavelength ultraviolet region (below 400 nanometres), the lifetime of the excited electronic state is sufficiently long that prior to the emission of radiation the molecule can undergo a series of vibrational state decays, lose energy through interstate transfer (intersystem crossing), or lose vibrational energy via molecular collisions.

Once a molecule arrives at the lowest vibrational level of an exited singlet state, it can do a number of things, one of which is to return to the ground state by photon emission. This process is called *°uorescence*. The lifetime of an excited singlet state is approximately

10¡9 to 10¡7 sec and therefore the decay time of °uorescence is of the same order of magnitude. If °uorescence is unperturbed by competing processes, the lifetime of °uorescence is the intrinsic lifetime of the excited singlet state. The *quantum e±ciency* of °uorescence is de¯ned as the fraction of molecules that will °uoresce.

It should be noted that even though a quantum of radiation is emitted in °uorescence this quantum will be lower in energy on the average than the quantum absorbed by the molecule, due to vibrational relaxation (both after absorption and after emission). The change in photon energy causes a shift of the °uorescence spectrum to longer wavelength, relative to the absorption spectrum, this is referred to as the Stokes Shift In addition to °uorescence, one also encounters *radiationless* processes where molecules in an excited singlet state may return to the ground state without the emission of a photon, converting all the excitation energy into heat.

The process called *internal conversion*, is not well understood and its e±ciency is very di±cult to measure. Generally internal conversion is an ine±cient process and is probably only a small fraction of the total excitation energy in most molecules. This is particularly true in aromatic hydrocarbons, which we use in this experiment. So far the discussion was limited to the $S0$! $S1$ transitions, and the question might arise as to the nature of exited state processes if a molecule is excited to a higher singlet state such as $S2$. In such cases the molecule will undergo vibrational relaxation as discussed before.

The fate of the molecule as it reaches the zeroth vibrational level of $S2$ depend of the energy separation between the exited singlet states. Generally the separation between excited singlet states in an aromatic molecule is smaller than the energy separation between the lowest singlet state $S1$ and the ground state $S0$. This means that the lowest vibrational level of $S2$ will overlap with higher vibrational levels of the $S1$ state, that do not involve extremely large di®erences in con¯guration from the zeroth vibrational level of the $S1$ state.

This situation gives rise to a high degree of coupling between the vibrational levels of the $S2$ and $S1$ states, which provides an extremely e±cient path for crossing from the $S2$ state to the $S1$ state. In fact this process is so e±cient that the molecule undergoes internal conversion from the $S2$ state to the lowest vibrational level of the $S1$ state in about the same time that it requires to convert from an exited vibrational level of the $S1$ state to its zeroth vibrational level (*i:e:* 10¡13 to 10¡11 sec). Because of this situation one may formulate the following rule: a molecule may be considered to undergo internal conversion to the lowest vibrational level of its lowest excited singlet state in a time that is short, relative to photon emission, regardless of the singlet state to which it was excited initially

Phosphorescence

Phosphorescence is a phenomenon exhibited in several fluorescent minerals where the mineral continues to glow after the UV light source has been removed. The glow slowly fades, and after several seconds (or minutes in a few cases) is no longer visible to the eye. Only few minerals phosphoresce; phosphorescence is only noted as interesting mineral property, rarely as a diagnostic identification property. Certain mineral specimens from specific localities all phosphoresce. Calcite specimens from Terlingua, Texas fluoresce bluish-white with a strong phosphorescence.

Phosphorescence is related to fluorescence in terms of its general mechanism but involves a slower decay. It occurs when a molecule whose normal ground state is a singlet is excited to a higher singlet state, goes to a vibrationally excited triplet state via either an intersystem crossing or a molecular collision, and subsequently, following vibrational relaxation, decays back to the singlet ground state by means of a forbidden transition. The result is the occurrence of a long lifetime for the excited triplet state; several seconds up to several hours are not uncommon. These long lifetimes can be related to interactions between the intrinsic (spin) magnetic moments of the electrons and magnetic moments resulting from the orbital motion of the electrons.

Molecules in singlet and triplet states react chemically in different manners. It is possible to affect chemical reactions by the transfer of electronic energy from one molecule to another in the reacting system. Thus the study of fluorescence and phosphorescence provides information related to chemical reactivity.

Although population of triplet states by direct absorption from the ground state is insigni¯cant, a more e±cient process exists for population of triplet states from the lowest excited singlet state in many molecules. This process is referred to as *intersystem crossing*, and is a spin-dependent internal conversion process. As singlet-triplet processes are generally less probable than singlet-singlet processes, one may be startled that a singlet-triplet process such as intersystem crossing can occur within the lifetime of an excited singlet state (10¡8 sec). The mechanism for intersystem crossing involve vibrational coupling between the excited singlet state and a triplet state.

Remembering that singlet-triplet processes are less probable than singlet singlet processes by a factor of 10¡5 to 10¡6, and that radiationless vibrational processes (such as internal conversion) occur in approximately 10¡13 sec, the time required for a spin-forbidden vibrational process would be approximately 10¡8 to 10¡7 sec, which is the same order of magnitude as the lifetime of an excited singlet state. Therefore intersystem crossing can compete with °uorescence emission from the zeroth vibrational level of an excited singlet state but cannot compete with vibrational deactivation from higher vibrational level of a singlet state.

Once intersystem crossing has occurred the molecule undergoes the usual internal conversion process (10¡13 to 10¡11 sec) and falls to the zeroth vi- brational level of the triplet state. Since the dierence in energy between the zeroth vibrational level of the triplet state and the zeroth vibrational level of the lowest exited singlet state is large compared to thermal energy, repopulation of a singlet state from a triplet state is highly improbable. There are two factors which tend to enhance a *radiationless* transition between the lowest triplet state and the ground state. First the energy dierence between the triplet state and the ground state is smaller than the dierence between the lowest singlet state and the ground state. This tends to enhance vibrational coupling between these two states, and therefore to enhance internal conversion. Second, and more important, the life time of a triplet state is much longer than that of an excited singlet state (about 10¡4 to 10 sec) and therefore loss of excitation energy by collisional transfer is generally enhanced. In fact, this second process is so important that in solution at room temperature it is often the dominant pathway for the loss of triplet state excitation energy. minimized, a *radiative transition* between the lowest triplet state and the ground state is observed. This emission is called *phosphore-scence.* As phosphorescence originates from the lowest triplet state, it will have a decay time approximately equal to the lifetime of the triplet state (ca. 10¡4 to 10 sec). Therefore phosphorescence is often characterized by an afterglow which is not observed for °uorescence.

Lasers and Masers

MASER

MASER stands for Microwave Amplification by Stimulation Emissionof Radiation. A LASER is a MASER that works with higher frequency photonsin the ultraviolet or visible light spectrum (photons are bundles of electromagnetic energy commonly thought of as "rays of light" which travel in oscillating waves of various wavelengths).

Maser is the abbreviation for "Microwave amplification by stimulated emission of radiation". Maser is an amplifier or generator for electromagnetic radiation usually microwaves. The laser is an optical maser. Einstein showed in 1917 how atoms, ions or molecules can emit radiation in the form of energy quanta (photons) through spontaneous (disordered photon emission) or photon emission stimulated through a signal.

It was unclear for a long time if stimulated emission happens orderly (amplify the signal) or if it also add together the photos disorderedly (increasing the noise). It wasn't until 1954 when C. H. Townes showed that it amplifies the signal.

The fundamental physical principle motivating the MASER is the concept of *stimulated emission*, first introduced by Einstein in 1917. Before defining it we look at two related but more familiar phenomena involving the interplay between matter and radiation, 0*absorption* and *spontaneous emission*.

Absorption: According to quantum mechanics, absorption of photons by atoms occurs only if the wavelength of the photon is just the right size (say, of wavelength l). If it is, the atom will "absorb" it (the photon vanishes) and go to a higher energy state. In physics, this process is called "absorption."

Spontaneous Emission: Atoms don't like to stay in high energy states (this is dictated by the laws of thermodynamics), so after absorbing a photon and going to a higher energy state, they will move of their own accord to a lower energy state, emitting a photon in the process. This is called "spontaneous emission" because no outside influence triggers the emission. Normally the average lifetime for spontaneous emissions by excited atoms is around 10^{-8} seconds (that is, the atom or molecule will usually take around 10^{-8} seconds before emitting the photon). Occasionally, however, there are states for which the lifetime is much longer, perhaps around 10^{-3} seconds. These states are called *metastable*. Metastable emission levels are essential for a working MASER and will be discussed further in a moment. Now that we've discussed absorption and spontaneous emission, we can get to stimulated emission (a MASER beam is made up entirely of stimulated emission).

Stimulated Emission: With stimulated emission, a photon of the absorption wavelength, l, is fired at an atom *already in its high energy state from prior absorption.* The atom absorbs this photon, and then quickly emits *two* photons to get back to its lower energy state. Thanks to quantum mechanics, both of these newly emitted photons are of wavelength:

Maser: In each frame, a molecule in the upper level of the MASER transition (that is, in the high energy, excited state) is indicated by a large red circle, while one in the lower level (low energy state) is indicated by a small blue circle.

- All of the molecules are in the upper state and a photon of wavelength l is incident from the left.
- The photon l stimulates emission from the first molecule, so there are now two photons of wavelength l, in phase.
- These photons stimulate emission from the next two molecules, resulting in four photons of wavelength l.
- The process continues with another doubling of the number of photons.

Basically, a man-made MASER is a device that sets up a series of atoms or molecules and excites them to generate the chain reaction, or amplification, of photons. Metastable emission states make MASERs and LASERs possible. To get the proper wavelengths to generate the chain reaction, first electricity or another energy source is "pumped" into a chamber filled with particular atoms or molecules. Then this "pumping" radiation causes the transition of atoms from the ground state to a high energy excited state higher than that referred to in the above paragraphs. From this short-lived state the atoms come down through non-radiative transition to the long-lived metastable state. Once in the metastable state many atoms can be accumulated in one place and in the same state.

Invention of Maser

Wartime work on electronics would lead to other discoveries besides solid-state electronic devices. Before the war, spectroscopy had generally focused on the high infrared, optical, and low ultraviolet regions of the EM spectrum, which could be picked up photographic film. That wasn't the case for the "microwave" region of the spectrum, the shortest wavelengths in the radio band; the technology didn't really exist to do a good job of mapping microwave spectral bands.

The war led to an explosive surge of development of microwave gear, and after the war physicists and chemists had a box full of new toys to probe the microwave region. Microwave spectroscopy came of age. Microwaves have much longer wavelengths than those found in the optical range, and accordingly much lower energy. While optical spectral lines are typically produced by the change of state of electrons in atoms or ions, microwave spectral lines are typically produced by changes in the configuration of molecular bonds. Microwave spectroscopy could be used to identify the "fingerprints" of various types of molecules, and also probe their behaviour.

Studies of the microwave spectra of molecules quickly convinced researchers that there was a significant gap in coverage, in the "millimeter wave" region of the electromagnetic spectrum, between microwaves and the low infrared. There hadn't been much work on millimeter wave technology during the war and so there weren't many tools available for their observation. The military was also interested in millimeter waves, since they could be used for advanced radars. Millimeter waves were easily blocked by raindrops and so on and didn't have much range, but the short wavelengths of millimeter waves meant that a millimeter-wave radar could provide a high-resolution image of a target.

In 1951, a physicist named Charles Townes from Columbia set up a conference of experts to consider the development of millimeter-wave technology. The technologies then available to produce microwaves weren't really up to the task, and Townes felt that something entirely new was needed. While he was considering the agenda for the meeting, an idea clicked together in his head.

Since some molecules could produce spectral lines in the millimeter-wave region of the spectrum, such molecules could be used to generate millimeter waves. It was hard to figure out an efficient way of doing this, though there was a phenomenon that helped. In 1917, while Albert Einstein was following up his thoughts about the photon theory of light, he discovered a concept known as "stimulated emission".

As discussed previously, if the electrons in an atom or molecule are excited to higher energy states, they will then tend to cascade back down to lower energy states, emitting a photon with an energy equal to that of the energy transition. Such emissions are generally "spontaneous", meaning that they happen at random times that follow a half-life curve, in a way very much like radioactive decay. Einstein realised that there was a way to provoke or "stimulate" such a emission. If an atom or molecule were hit by a photon with an energy matching that of a difference between two particular energy states and the electron was in the higher energy state, it would fall down to the lower energy state and release a photon of the same energy as the one that hit the atom or molecule in the first place. In addition, the two photons would be "coherent", moving in the same direction and having the same phase.

In 1950, the researchers, had achieved a population inversion in a crystal of lithium fluoride as a "pure physics" experiment, demonstrating that a population inversion was possible. As for the limitations of stimulated emission, Townes thought he could perform it in an electromagnetic resonant cavity, a chamber precisely machine to allow electromagnetic radiation of a specific wavelength to resonate and build up as the beam bounced from one end of the cavity to the other, stimulating emission from the molecules with each pass.

By 1951, the team had come up with a scheme for generating radiation using ammonia (NH3) molecules, with output at a wavelength of 1.25 centimeters. This was not millimeter-wave emission, but Townes and his group had found that generating millimeter waves was very difficult, partly because of the chicken-and-egg problem of a lack of millimeter-wave components. The group was trying to perform a proof-of-concept demonstration with microwaves and then move on to shorter wavelengths.

The NH3 molecule is a tetrahedron, with the nitrogen atom linked to a base of three hydrogen atoms connected in a triangle. In its ground state, the molecule is polarized, with a negative charge on the nitrogen end and a positive charge on the hydrogen base of the tetrahedron. Such a polarized molecule could be moved by an electric field. However, when an ammonia molecule is excited, the nitrogen atom has an equal probability of being on either side of the base, with the result that the molecule is no longer polarized and won't move in an electric field. That meant that ammonia molecules could be heated and then fed into a resonant cavity through an assembly that set up an electric field to shunt the ground-state molecules off to one side. By 1953, the researchers were able to use the scheme to amplify microwaves, and were using it as a microwave oscillator. The device was named a "maser", for "microwave amplification by stimulated emission of radiation". The term "amplification" was used instead of "oscillation" because the maser ended up being a very sensitive low-noise amplifier, often used in radio astronomy because it allowed pickup of faint cosmic radio signals. Ironically, radio telescope observations would also show that maser action occurred in nature, with stimulated emission in cosmic molecular gas clouds creating a relatively intense emission.

Laser

Laser is the abbreviation for “Light amplification by stimulated emission of radiation”. Laser light is emitted when many atoms undergo similar energy transitions at the same time. This is achieved by promoting a large number of atoms to an energy level above the ground state. As an electron in one of the excited atoms jumps down from its higher energy level it emits a photon. As this photon travels past another atom in an excited state, it causes the electron in this atom to jump down to the lower level. The passage of light thus stimulates the emission of radiation from other atoms – producing the intense beam of light characteristic of the laser.

The laser was first demonstrated by the American Theodore Maiman in 1960 using a ruby laser. The stimulated emission of radiation was initially postulated by Einstein in 1917. The laser’s counterpart in the microwave part of the spectrum paved the way for the production of laser light after two American scientists Schawlow and Townes made a theoretical paper proposing how the maser technology could be widened to fit the visible part of the spectrum.

A laser consists of an active medium which is placed between two mirrors. The arrangement of the mirrors is called a laser cavity. On of the mirrors is semi-transparent to release the laser that is generated in the cavity. Energy must be supplied from ex. flash bulb.

The supplement of energy in the form of a for instance flash bulb is called pumping. The active medium consists of atoms or molecules. Normally almost all of the atoms are in their lowest energy level, the so called ground state. From there they can be transferred to a higher, excited energy level through absorption of a light quantum (a photon) with the energy “E = hv. The upper state is often very short-lived (microsecond to nanosecond) and the atom returns to the ground state during the emission of radiation. Two radiation processes can happen; spontaneous emission or stimulated emission. The spontaneous emission consist of a random emission of photons in random direction is responsible for the excited state being so short-lived. The stimulated emission can happen if the atom is shone with radiation that has the frequency that is equal to the transition (v). The stimulated photons are emitted with the same frequency as that of the stimulating photons. The probability for the decay of an excited atom is equal to the probability for absorption of a photon by an atom in its ground state and being excited. The stimulated emission can be regarded as a negative absorption.

Normally a beam of light is weakened by a medium when the atoms in it undergo energy level transitions. If you on the other hand get more atoms in the higher energy level that in the lower (population inversion) the amplification of the beam through stimulated transitions from the higher energy level will dominate over the weakening of the beam through the absorption of light of the atoms in the ground state. The result is amplification by stimulated emission of radiation.

The two mirrors form a resonator which is an apparatus for identification or amplification of a special frequency around the amplifying active medium. As long as the population inversion stands, there will be an enormous amplification along the passage between the mirrors. For each and every atom that returns to the ground state, the beam gains another photon. The laser is reflected back and forth in the cavity until it has enough intensity to be sent out of the cavity. After that, the flash bulb can again excite enough atoms for the creation of a new pulse. This is the principle behind a pulsating laser. In a continued laser, the laser light is released all the time. If an obstacle is but in the laser cavity the laser discharge is stopped. If the obstacle is removed when the flash bulb has excited a large number of atoms it becomes one gigantic pulse with a today maximum effect of a few terawatt. Lasers can be modified to have an enormous sharp in frequency, effect in pulse and length of pulse. This can't be obtained in the same laser system. According to Heisenberg's uncertainty relationship an ultra short pulse can not be sharp in frequency.

Invention of Laser

Although the maser was ingenious and won the admiration of the physics community, from the public point of view it attracted little or no attention. It was a useful research tool, but had few or no consumer, industrial, or military applications. However, Townes had no intention of stopping at the maser, and in fact he had basically developed it as a stepping stone to a millimeter-wave system. After spending a few years getting maser technology to work reliably, in 1957 he went back to the original problem of generating millimeter waves. Much to his surprise, his initial theoretical analysis towards this end showed that it would be about as practical to generate infrared radiation or visible light using stimulated emission in a resonant cavity as it would be to generate millimeter waves. This was a bit of a shock, since over the previous decades every order-of-magnitude reduction in the operating wavelengths of radio gear had taken the development of a new generation of technology. Now Townes saw the possibility of jumping over five orders of magnitude in a single bound; he rechecked his figures and decided that it really was possible.

He joined forces with a Canadian physicist named Arthur W. Schawlow, who worked at Bell Laboratories and also happened to be Townes' brother-in-law. They came up with a notion of using a lighting system to excite potassium gas in a long cylindrical tube with a mirror at one end and a half-silvered mirror at the other in which the light could build up; such a structure became known as a "Fabry-Perot resonator". In the summer of 1958, they applied for a patent on the concept and then published their ideas in the physics press. Other people were thinking along the same lines. Robert H. Dicke of Princeton wrote up some notes on the idea in 1956 but didn't follow them up. A graduate researcher at Columbia named R. Gordon Gould also came up with the notion in 1957; he had dealings with Townes and suspected Townes was likely working along the same lines. Gould wrote up some notes and had them notarized in late 1957, with the notes suggesting applications such as optical radars, optical communications, and even laser-driven fusion.

After the publication of their paper, Townes and Schawlow more or less went their separate ways; townes worked on the potassium-gas optical maser, while Schawlow focused on a crystal of synthetic pink ruby. Pink ruby is a crystal of aluminum oxide, with traces of chromium atoms embedded in the crystal matrix; the chromium atoms would provide the stimulated emission. However, Schawlow decided that pink ruby wasn't very promising and gave it up. Another Bell Labs researcher, an Iranian-born physicist named Ali Javan, had some acquaintance with Townes and Schawlow and had read their 1958 paper. Javan had other ideas for an optical maser, believing that he could build one using an electrical discharge tube containing helium and neon and arranged as a Fabry-Perot interferometer. Following a conference on quantum electronics in late 1959, a Hughes researcher named Theodore H. Maiman jumped into the race, focusing his work on an optical maser using a pink ruby crystal.

Different Types of Laser

Gas Laser:

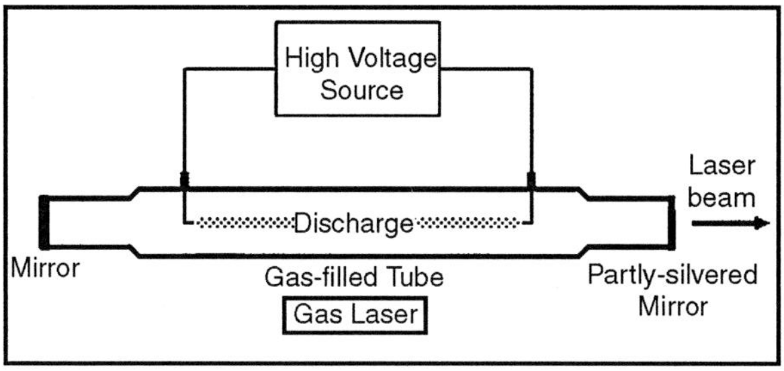

Common gas laser types include:

- Helium-neon lasers, which normally produce red light but apparently can also produce green light. As mentioned, Ali Javan's pioneering gas laser was a helium-neon laser.
- Argon lasers, which produce blue-green light.
- Carbon dioxide lasers, which produce infrared light. CO2 lasers can provide outputs of up to about 100 kW, and are used in industrial cutting and welding.
- Carbon monoxide lasers, which are very powerful, with output power of up to 500 kW. Not surprisingly, they require cooling.
- "Excimer" gas lasers, which produce short-wavelength ultraviolet light, making them useful for applications where a very precise beam is required, such as semiconductor manufacturing or eye surgery. The gases are compounds of fluorine or chlorine such as F2, ArF, KrCl, KrF, XeCl, and XeF. They are known as "excimers" for "excited dimers", because the molecules are formed and then broken apart to generate their high-energy ultraviolet photons.
- Metal vapour lasers, featuring vaporized metals such as cadmium in a buffer gas of helium.

Crystalline and Glass Lasers

Crystalline and glass lasers are usually optically pumped by a flashlamp, LEDs, or laser diodes. The original red-light ruby laser is still in use, but a variant of garnet known as "yttrium-aluminum-garnet (YAG)" may be doped with rare earths such as neodymium, erbium, thulium, ytterbium, or holmium to and operate in various subbands in the infrared. Other examples include titanium sapphire lasers, which can be tuned to generate very short pulses, and high-power neodymium glass lasers. There are some interesting variations on crystalline and gas lasers. One involves doping a section of a fibre-optic thread with the rare earth erbium; this scheme is used to boost laser signals being sent over fibre-optic communications links. Another involves using dyes in a liquid substrate, with the dyes selected to give highly specific operating wavelengths. Such "organic dye lasers" were invented in 1966 by John R. Lankard and Peter Sorokin.

The semiconductor laser was first suggested by Basov in the late 1950s. The first semiconductor lasers were simple PN-junction devices. They worked, but they were current hogs and were not suitable for anything but lab tests. The problem was that the junction region in which lasing activity took place was simply too big: even though it was on the order of a micrometer in depth, that still made probability of stimulated emission too small.

The advance that made semiconductor lasers possible was the "double heterojunction (DH)" laser, developed in the early 1970s. It featured a thin gallium arsenide (GaAs) layer about 0.2 microns thick sandwiched between p-type and n-type aluminum gallium arsenide (AlGaAs), with the thin layer providing a much higher probability of stimulated emission, and permitting laser action at reasonable current levels.

Diode Laser

Laser diodes seemed ideal for transmitting digital data over a fibre optic link, with their crisp pulses permitting higher data rates than LEDs, but early laser diodes didn't generate light of the right wavelength to pass through a fibre optic thread with minimal attenuation. Changing the wavelength meant tweaking with the semiconductor material, using alternative alloys such as aluminum gallium arsenide antimony (AlGaAsSb) and, particularly, aluminum gallium arsenide phosphide (AlGaAsP).

The DH laser made the laser diode practical, but it still left much to be desired. Although laser action was confined to a thin layer, it was not very well confined*within* that thin layer. Fabricating a laser diode as simple layers of materials wasn't good enough. The next step, the "buried heterojunction (BH)" laser, was fabricated using new semiconductor etching techniques to create the lasing region as a "wedge" crammed between nonconductive materials on each side that provided "lateral confinement". Several other device configurations were also developed to provide lateral confinement.

Semiconductor lasers

Semiconductor lasers continue to be refined. One modern device includes a "control gate" element, in which a current adjusts the refractive index of the device's lasing channel, which adjusts the wavelength of the output. Such devices can be rapidly switched between wavelengths. Another interesting modern device is the "vertical cavity surface emitting laser (VCSEL)", in which lasing takes place between the top and bottom of the device, not between the sides as with traditional laser diodes. VCSELs can be built as arrays containing large numbers of tiny lasers in a grid, which can be used in optical image processing systems.

Although silicon is not inherently a very good material for building semiconductor lasers, there is still interest in silicon lasers because they could be integrated on the same chip with silicon microelectronics, not only potentially reducing cost of the overall system but providing such niceties as fast optical internal links for silicon microchips. For the moment, silicon lasers remain lab toys. The earliest had to be activated by a second laser, which made them of minimal usefulness, but improved devices have appeared that can be electrically driven to generate a laser beam. These devices manage to get around the limits of silicon by employing the "Raman effect", a shift in wavelength that occurs when a photon interacts with an atom in a transparent material. Free electrons in silicon tend to damp laser action, but the experimental devices use a matrix of electrodes that pull electrons out of the lasing channel.

Semiconductor lasers are traditionally low-power devices, useful for consumer and communications applications, though now there are industrial semiconductor lasers suitable for cutting and welding.

Laser Applications

Gyroscopes: One of the important applications of the laser in aerospace applications is in the form of the "optical gyroscope" or "laser gyroscope", used as the core element of an "inertial navigation system (INS)". An INS is carried by an aircraft or other vehicle to give its position relative to a point of origin. In modern times, navigation is primarily performed using the GPS navigation satellite system, but since GPS can be jammed, a GPS navigation system usually has an INS backup.

Early INS units developed in the 1950s used mechanical gyroscopes to determine changes in direction. Although such technology was highly refined, ultimately featuring mechanical gyros spinning on a film of gas to reduce friction that could throw off their accuracy, there was no way to eliminate friction entirely, limiting the accuracy of the INS.

The development of the laser led to a new approach, the optical gyroscope, in which a rotating loop of light was used to determine changes in direction. The optical gyroscope was not affected by friction. The basic principle was not new, going back to a paper published by the French physicist Georges Sagnac before the First World War, but the concept wasn't practical until the introduction of the laser.

An optical gyroscope is a type of optical interferometer, in which two laser beams travel in opposite directions, or "counterpropagate", over a closed path, or "optical cavity", inside the gyroscope. The two beams optically interfere with each other, and the effects of motion on the interference can be used to determine the rate and direction of rotation of the system. The first optical gyroscope, the "ring laser gyro (RLG)", was developed in 1962 by Warren Macek of Sperry Corporation, and the device was quickly incorporated into the guidance systems of missiles and other vehicles. The classic RLG is built out of a flat triangular block of "Zerodur", a type of glass that resists changes in dimensions with temperature, with a channel fabricated into it not only to act as a resonant path for the gyro's laser beam, but also to generate the laser beam itself. This is why the classic RLG is sometimes called an "active optical resonator".

The channel is filled with a lasing gas mix, usually helium-neon. Three highly reflective mirrors are placed at the squared-off corners of the triangular block. Anodes are placed on two sides of the block and a cathode is place on the third to provide laser stimulation. One of the corner mirrors is semi-transparent, to allow reading the laser light into a detector. In effect, a classic RLG is a helium-neon laser bent into a loop.

Consider the effect of rotation on the length of the path the light beams traverse in the resonant cavity. If the tube rotates in one direction, the light beam propagating in that direction sees a longer path length, while the one propagating in the other direction sees a shorter path length. This is the "Sagnac effect".

The frequency of laser light is dependent on two factors: the spontaneous light emission frequency of the lasing material, and to a lesser extent the cavity length. Lasing only occurs if there are an integral number of wavelengths of light in a cavity, and so changing this length can shift the laser's frequency from the centre frequency of the lasing medium, at least up to the point where the cavity becomes an integer multiple of that centre frequency again.

Under rotation, the Sagnac effect changes the path lengths seen by the counterpropagating light beams, resulting in raising the frequency of the beam moving against the rotation and lowering the frequency of the beam moving with the rotation. This frequency difference results in optical interference effects that can be measured by the output detector.

The laser medium is excited by a DC current, which can lead to gas flows that throw off the lasing centre frequency through Doppler shifts. This is why the RLG has two anodes: they set up flows in opposite directions, and their effects cancel.

The RLG also is subject to a troublesome problem known as "frequency lock-in". At low rotation rates, a small amount of "backscatter" from the reflective lasing mirrors interacts with the lasing medium and causes the counter propagating beams to stay locked together, and the output does not change until the rotation increases beyond a certain threshold, resulting in a "dead band" in the output range. The simplest technique for reducing the dead band is "mechanical dithering", in which the RLG is vibrated over a small arc at a typical frequency of 400 hertz. Dithering tends to lead to unstable output if the rates of rotation are high, as they often are in high-speed aircraft or spacecraft. Some researchers have experimented with RLGs using pulsed lasers as a potential solution.

Frequency lock-in won't occur if the laser beam is sourced outside the ring cavity, and experiments have been performed with RLGs using an external laser source. Such devices were generally relatively bulky, complicated, and expensive, and their advantages were not sufficient to replace the classic RLG.

Refinements in fibre-optic technology for communications led to the "fibre-optic gyroscope (FOG)", which uses a coil of fibre-optic thread for the laser loop. The FOG offers a much cheaper and more compact alternative to the RLG, but still has good performance. A FOG requires an external laser source, generally a LED or laser diode.

Laser Radar and Communications

Another early application of lasers was for range finding. Radar had been used for this purpose since World War II, using the simple principle of sending out a pulse of radio waves and measuring the time until an echo to return from a target. Multiplying half the time by 300,000 gave the range in kilometers.

The accuracy of radar ranging is a function of the wavelength and pulse length, with accuracy improving as both get shorter. However, sensitivity of radar is a function of total pulse power, which increases with pulse width. Lasers could provide very short pulses with short wavelengths and high power, offering the best of both worlds. Laser rangefinders were in use in military applications by the 1970s; since a laser rangefinder required a laser detector system, some aircraft used laser rangefinders that also provided a laser spot tracker capability. Modern infantry laser target designators may include a laser range finding capability. The latest laser rangefinders/ designators may also have a GPS receiver and a communications interface, allowing them to determine the location of a target and then relay its coordinates to a "strike platform" for attack by a GPS-guided weapon.

Laser rangefinders have scientific applications as well. The US Mars Global Surveyor spacecraft used a precision laser altimeter to obtain a highly accurate map of Martian ground levels. Earth-based laser rangefinders have been used to target reflector units left on the Moon by the Apollo manned Moon missions to determine the rate at which the Moon is receding from the Earth and the rates of continental drift.

Experimental imaging laser radar systems have been developed. They provide very high resolution compared to microwave imaging radar systems, but are limited in range by dust or fog. Yet another application of laser ranging is in laser 3D scanners, in which a laser beam is precisely scanned over an object to digitize it into a set of three-dimensional digital coordinates, possibly for generating 3D computer graphics imagery.

Lasers have become extremely important for high-speed digital communications. The amount of information that can be crammed into a communications channel is in principle proportional to the frequency, and the high frequencies/ short wavelengths of light mean that a data link using light can carry much more information than a data link using longer wavelength radio waves.

The biggest problem with using light for data transmission is of course the fact that it is easily blocked by rain or dust in the air, but the introduction of fibre-optic communications cables in the 1970s provided an uninsured long-distance path. LEDs were used for optical data links at first, but semiconductor lasers could generate shorter and cleaner pulses, permitting higher data transmission rates, and they now predominate. Since there's little in space to interfere with a laser beam, optical communications links are being considered for spacecraft. Such "free space" optical links would allow spacecraft observing Mars or other planets to send large volumes of data back to Earth, to be picked up by satellites with telescopic optical receivers. Some researchers interested in searching for extra-terrestial (ET) civilizations have considered the possibility that ETs are using powerful lasers to communicate over interstellar distances and are searching for such laser emissions.

More practically, blue-green lasers, which can penetrate seawater to a substantial depth, have been used to communicate with underwater submarines, and incidentally as elements of sensor systems used to hunt naval mines.

Laser Energy Weapons and Fusion Power

Of course, the original vision of a laser as an offensive weapon, a "death ray", has not died out, and in fact seems not too far away. As a first step, lasers are now being used in directed infrared countermeasures (DIRCM)" systems carried by aircraft to dazzle heat-seeking missiles and cause them to go astray. There was some talk of using lasers to blind enemy troops as well, but the public reaction to such concepts has been loud and negative.

The next step is to use a laser to destroy the missiles, not simply dazzle them. The US military has developed prototype diode-pumped battlefield lasers to intercept small rockets and even artillery shells. Such systems are not ready for fielding yet, but progress is rapid at this time, and attack laser systems for aircraft or armored vehicles are expected to be available in the 2010:2020 timeframe.

The Air Force is now developing a very high power chemical laser that will be shot out of a modified Boeing 747 jumbo jet to intercept missiles just after tey are launched. The Airborne Laser (ABL)" uses a "combustive oxygen and iodine laser (COIL)" system based on light emission in the burning of oxygen and iodine; earlier exercises along this line used hydrogen fluoride lasers, based on the combustion of ethylene (C2H4) and nitrogen trifluoride (NF3), the fluorine analogue of methane. The Air Force is also interested in developing a satellite platform with a high energy laser to shoot down missiles. During the "Strategic Defence Initiative (SDI)" or "Star Wars" effort of the 1980s, work was done on an X-ray laser system that, it was hoped, could shoot down missiles as well. A coherent X-ray beam would be very energetic and implementing shielding against it would be very difficult.

The concept involved a laser system that was pumped by the detonation of a small nuclear weapon, generating the beam in the instant before the weapon destroyed itself. However, demonstration tests of the scheme did not produce useful results and it was abandoned.

In an inverse application, lasers have been used to experimentally produce nuclear reactions, if on a small scale. As mentioned earlier, high-power laser systems have been used to perform fusion in deutrium/ tritium fuel pellets as a means of performing controlled fusion. Initially, physicists were very optimistic about laser fusion, believing that it would only take modest laser power to do the job. Unfortunately, the plasma formed around the pellet during the process had instabilities that required laser power an order of magnitude greater than expected.

It didn't turn out to be so simple after all. The technology is not remotely close to producing a practical fusion reactor, but the lasers used in the experiments have been impressive. The Nova laser, at Lawrence Livermore National Laboratories is one of the world's biggest, a neodymium-glass apparatus with ten arms that covers an area about the size of an athletic field and can focus a 100 terawatt pulse onto a strip of selenium, which produces an X-ray beam that is focused on a capsule of deuterium.

The French Atomic Energy Commission, in hopes of demonstrating the feasibility of inertial confinement fusion, is now underway on the National Ignition Facility (NIF) at Lawrence Livermore Labs, and the Laser Megajoule (LMJ) facility near Bordeaux, building development. The NIF laser system will be 60 times more powerful than the Nova laser system. A small group of advocates has been promoting a less brute-force scheme, known as "fast ignition", which uses two relatively low-power lasers instead of one big laser system. In fast ignition fusion, one laser keeps the fuel pellet confined, while the second laser is focused into the pellet through a gold cone to drive fusion. Even the advocates admit that fast ignition is highly theoretical at this time and needs a lot of proving.

Chapter 7

Stereochemistry

Stereochemistry, a subdiscipline of chemistry, involves the study of the relative spatial arrangement of atoms within molecules. An important branch of stereochemistry is the study of chiral molecules. Stereochemistry is a hugely important facet of chemistry and the study of stereochemical problems spans the entire range of organic, inorganic, biological, physical and supramolecular chemistries. Stereochemistry includes methods for determining and describing these relationships; the effect on the physical or biological properties these relationships impart upon the molecules in question, and the manner in which these relationships influence the reactivity of the molecules in question (dynamic stereochemistry).

Louis Pasteur could rightly be described as the first stereochemist, having observed in 1849 that salts of tartaric acid collected from wine production vessels could rotate plane polarized light, but that salts from other sources did not. This property, the only physical property in which the two types of tartrate salts differed, is due to optical isomerism. In 1874, Jacobus Henricus van 't Hoff and Joseph Le Bel explained optical activity in terms of the tetrahedral arrangement of the atoms bound to carbon.

One of the most infamous demonstrations of the significance of stereochemistry was the thalidomide disaster. Thalidomide is a drug, first prepared in 1957 in Germany, prescribed for treating morning sickness in pregnant women. The drug however was discovered to cause deformation in babies. It was discovered that one optical isomer of the drug was safe while the other had teratogenic effects, causing serious genetic damage to early embryonic growth and development. In the human body, thalidomide undergoes racemization: even if only one of the two stereoisomers is ingested, the other one is produced. Thalidomide is currently used as a treatment for leprosy and must be used with contraceptives in women to prevent pregnancy-related deformations. This disaster was a driving force behind requiring strict testing of drugs before making them available to the public. Cahn-Ingold-Prelog priority rules are part of a system for describing a molecule's stereochemistry. They rank the atoms around a stereocenter in a standard way, allowing the relative position of these atoms in the molecule to be described unambiguously. A Fischer projection is a simplified way to depict the stereochemistry around a stereocenter.

Atropisomer

Atropisomers are stereoisomers resulting from hindered rotation about single bonds where the steric strain barrier to rotation is high enough to allow for the isolation of the conformers. The word atropisomer is derived from the Greek *a* which means *not* and *tropos* which translates as *turn.*

The name was coined by Kuhn in 1933, but atropisomerism was first detected in 6,6'-dinitro-2,2'-diphenic acid by Cristie in 1922. Oki defined atropisomers as conformers that interconvert with a half-life of more than 1000 seconds at a given temperature.Atropisomers are an important class of compounds because they display axial chirality. They differ from other chiral compounds in that they can be equilibrated thermally whereas in the other forms of chirality isomerization is usually only possible chemically.

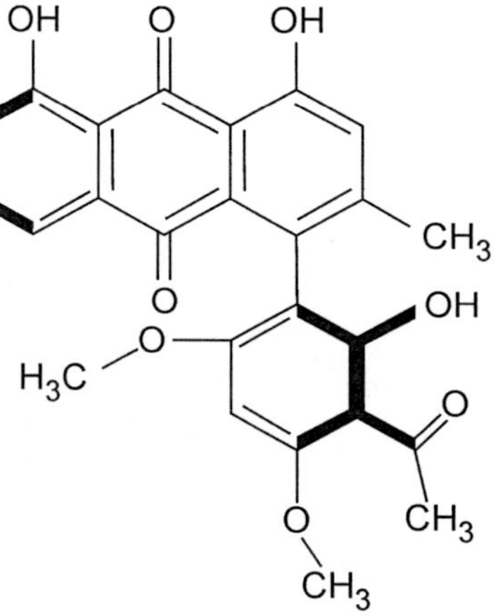

The most important class of atropisomers are biaryls such as the aforementioned diphenic acid which is a derivative of biphenyl with a complete set of ortho substituents. Others are dimers of naphthalene derivatives such as 1,1'-bi-2-naphthol. In a similar way aliphatic ring systems like cyclohexanes linked through a single bond may display atropisomerism provided that bulky substituents are present. Examples of naturally occurring atropisomers include vancomycin and knipholone which is found in the roots of Kniphofia Foliosa of the family Asphodelaceae.

Separation of atropisomers is possibly by chiral resolution methods such as selective crystallization. In an atropo-enantioselective or atropselective synthesis one atropisomer is formed at the expense of the other. Atroposelective synthesis may be carried out by use of chiral auxiliaries like a CBS catalyst in the total synthesis of knipholone or by approaches based on thermodynamic equilibration when an isomerization reaction favors one atropisomer over the other.

Scope

In one application the asymmetry in an atropisomer is transferred in a chemical reaction to a new stereocenter. The atropisomer is an iodoaryl compound synthesised starting from (S)-valine and exists as the (M,S) isomer and the (P,S) isomer. The interconversion barrier between the two is 24.3 kcal/mol (101.7 kJ/mol).

The (M,S) isomer can be obtained exclusively from this mixture by recrystallisation from hexanes. The iodine group is homolytically removed to form an aryl radical by a tributyltin hydride/triethylboron/ oxygen mixture as in the Barton-McCombie reaction.

Although the hindered rotation is now removed in the aryl radical the intermolecular reaction with the alkene is so much faster than rotation of the carbon-nitrogen bond that the stereochemistry is preserved. In this way the (M,S) isomer yields the (S,S) dihydroindolone.

An axial chirality switch is reported for a diol prepared from intramolecular pinacol coupling of the corresponding di-aldehyde with samarium(II) iodide. In methanol this compound has the two alcohol groups in equatorial positions but in hexane helicity is reversed with both groups in axial positions.

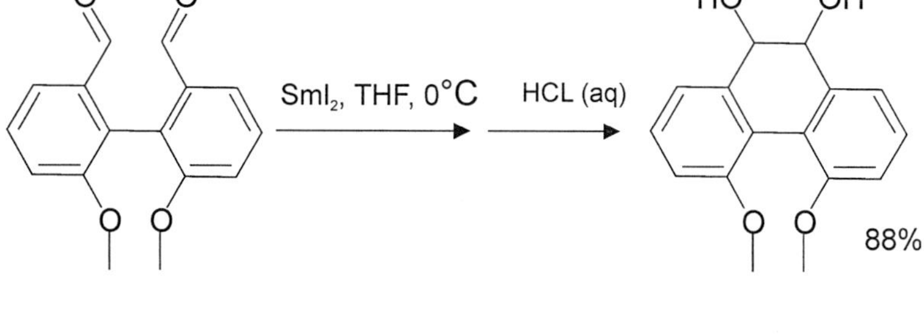

Asymmetric Catalysis

In organic chemistry BINAP is a ligand that is used in the preparation of optically active stereoisomers.

Conformational Isomerism

In chemistry, conformational isomerism is a form of stereoisomerism in which molecules with the same structural formula (same connectivity) exist as different conformational isomers or conformers in 3-D due to rotations about one or more σ bonds. Conformations are not superimposable in three dimensions and can be characterized by their dihedral angles, the angle between two planes formed by $atom_1$-$atom_2$-$atom_3$ and $atom_2$-$atom_3$-$atom_4$ in a sequence of atoms, $atom_1$-$atom_2$-$atom_3$-$atom_4$. The dihedral angle can also be seen as the angle between the two vectors formed by $atom_2 \rightarrow atom_1$ and $atom_3 \rightarrow atom_4$. Rotamers are conformers that differ by rotation about only a single σ bond. That is, only one dihedral angle will be different between rotamers. Conformers can differ by rotations about many sigma bonds or just one (i.e. a rotamer is a conformer). The rotational barrier, or barrier to rotation, is the activation energy required to convert from one rotamer to another rotamer.

Different conformers can interconvert by rotation around single bonds, without breaking chemical bonds. The existence of more than one conformation, usually with different energies, is due to hindered rotation about sp^3 hybridised σ bonds. The comparative stabilities of different conformers of a molecule are usually explained through differences in a combination of steric repulsion and electronic effects. A simplified example is that of a butane molecule viewed in the Newman projection shown-i.e. as if viewed down the central C2-C3 bond with relative rotations of C1 and C4 illustrated. The only unique gauche conformer in case of butane has a dihedral angle of 60°, anti is 180°, and eclipsed is 0° in the CH_3-CH_2-CH_2-CH_3. In the case of 1-fluoropropane F-CH_2-CH_2-CH_3, there is a gauche$^+$ and a gauche$^-$ rotamer, with dihedral angles of +60 and -60 respectively. Butane does not have both because of symmetry.

Important examples of conformational isomerism include:

1. Linear alkane conformations with staggered, eclipsed and gauche conformers, and
2. Ring conformation
   — Cyclohexane conformations with chair and boat conformers.

— Carbohydrates
— Steroids

3. Atropisomerism- due to restricted rotation about a bond, a molecule can become chiral
4. Folding of molecules, where some shapes are stable and functional, but others are not.

Consequences

If the eclipsed conformations of an isomer have high enough potentials, they may prevent rotation of substituents to different staggered conformations at sufficiently low energy levels. This will result in a racemic mixture of conformations that may or may not have different reactivities in situations such as enzymatic reactions in which molecular shape is usually a key factor of operation.

Conformer Dependent Reactions

The E2 elimination mechanism relies on the base- or acid-attacked substituent being in an antiperiplanar configuration along a bond with respect to the leaving group. This prerequisite for reaction is important in understanding organic elimination reaction pathways, especially those involving halogenated cyclic alkanes such as cyclohexanes. Two adjacent substituents on a cyclic alkane can only undergo an E2 elimination if they are both axial to the ring and hence antiperiplanar. A combination of axial and equatorial substituents cannot react through an E2 mechanism, though ring flips (with associated reconformation) may allow reactions to occur if they are not precluded by an energy barrier or steric lock through isopropyl or larger substituents.

Conditions

Conformational isomerism only occurs around single bonds because double or triple bonds have one or two pi bonds that prevent rotation about the longitudinal axis. Conformers sufficiently constrained to exhibit measurable isomerism are unique from various flavours of stereoisomers in the fact that changes in stereochemistry are independent from any mechanism and instead rely only on molecular energy.

Basic Concepts of Stereochemistry

Isomers are molecules that have the same molecular formula but different arrangements of their constituent atoms. *Constitutional isomers* (*structural isomers*) have different bonding arrangements of their atoms (*connectivity*) and usually show very marked differences in physical and chemical properties. Connectivity differences can involve the carbon skeleton or the nature and position of functional groups.

The Cahn–Ingold and Prelog Sequence Rules

In order to categorise stereoisomers it is necessary to prioritise different atomic substituents using the *Cahn–Ingold and Prelog sequence rules* that are applied in order until a distinction is found. Rank substituents in order of decreasing atomic number of the first bound atom. Any higher isotope takes precedence over a lower isotope ($H^3 > H^2 > H^1$). A lone pair counts as the lowest priority substituent. If no distinction is possible at the first atom, consider atoms at increasing distances until a difference is found. For the purposes of ranking, a multiple–bonded atom is considered equivalent to the same multiple of single–bonded atoms and has higher priority than the corresponding single–bonded substituent.

Enantiomers

An *enantiomer* is one of a pair of stereoisomers that are related as non–superimposable mirror images. Enantiomerism commonly results from the presence of one or more stereogenic centres in a molecule but may also occur in orthogonal structures (allenes, hindered biaryls), helical structures (*E*–cyclic alkenes, helicenes) and extended tetrahedra (differentially substituted adamantanes). Such molecules are *chiral* and display identical chemical and physical properties in an achiral environment.

However, opposite enantiomers will react at different rates with a single enantiomer of a reagent. A solution of a single enantiomer will rotate the plane of plane–polarised light and is referred to as *optically active*; although this physical property cannot be directly related to absolute configuration of the molecule. An enantiomer is given the prefix (+)– if the rotation is clockwise (*dextrorotatory*) and (–)– if the rotation is anticlockwise (*levorotatory*).

Bibliography

Amidon, G. L.: *Chemical Stability of Pharmaceuticals*, Wiley-Interscience, New York, 1986.

Atkins, Peter and Friedman, Ronald: *Molecular Quantum Mechanics*, Oxford University Press, New York, 2005.

Banki, L.: *Bioassay of Pesticides in the Laboratory,* Akademiai Kiado, Budapest, 1978.

Bard, A.J.; Faulkner, L.R.: *Electrochemical Methods: Fundamentals and Applications*, John Wiley & Sons, New York, 2005.

Barnum, Susan R.: *Biotechnology: An Introduction*, Belmont, Thomson/Brooks/Cole, 2005.

Barrington, E. J. W.: *Biochemistry of Primitive Deuterostomians*, London, Academic Press, 1974.

Bell, Eric T.: *The Magic of Numbers*, New York, McGraw-Hill, 1946.

Benvenuto, E.: *An Introduction to the History of Structural Mechanics*, Springer-Verlag, New York, 1991.

Bhatnagar, Vasudev : *Cell Science and Technology*, Campus Books, Delhi, 2009.

Bhattacharya, Lata: *Biochemistry of Nutrition*, Discovery, Delhi, 2010.

Boyle, Robert: *The Sceptical Chymist*. Dover Publications, New York, 1661.

Brinker, C.J.; G.W. Scherer: *Sol-Gel Science: The Physics and Chemistry of Sol-Gel Processing*. Academic Press, Boston, 1990.

Brown, H. C., D. H. McDaniel, and O. Hauiger: *Determination of Organic Structures by Physical Methods*, Academic Press, New York, 1955.

Bummer, P. M.: *The Science and Practice of Pharmacy*, Mack, Easton, PA, 2000.

Burns, G.; Glazer, A.M.: *Space Groups for Scientists and Engineers,* Academic Press, Boston, 1990.

Burstall, A. F.: *A History of Mechanical Engineering*, Faber and Faber, London, 1963.

Callen, H. B.: *Thermodynamics*, Wiley, New York, 1960.

Campbell, Neil A.; Brad Williamson; Robin J. Heyden: *Biology: Exploring Life*. Pearson Prentice Hall, Boston, Massachusetts, 2006.

Carstensen, J. T.: *Pharmaceutics of Solids and Solid Dosage Forms*, Wiley-Interscience, New York, 1977.

Chandler, David: *Introduction to Modern Statistical Mechanics*, Oxford University Press, New York, 1987.

Chartres, C. and Varma, S.: *Out of Water, From Abundance to Scarcity and How to Solve the World's Water Problems,* FT Press, USA, 2011.

Clayden, J., Warren, S.: *Organic Chemistry*, Oxford University Press, Oxford, 2000.

Cohen, Lizabeth, *Making a New Deal, Industrial Workers in Chicago, 1919-1939* Cambridge University Press, 1991.

Crouch, Stanley; Skoog, Douglas A.: *Principles of Instrumental Analysis*, Thomson Brooks/Cole, Australia, 2007.

Cullity, B.D.: *Elements of X-Ray Diffraction*, Reading, Addison-Wesley Publishing Company, Massachusetts, 1978.

Davenport, Thomas: *Process Innovation: Reengineering Work through Information Technology*, Harvard Business School Press, Boston, 1993.

David, M.: *Ideas in Chemistry: A History of the Science,* New Brunswick, N.J.: Rutgers University Press, 1992.

De Gennes, Pierre Gilles: *Scaling Concepts in Polymer Physics*, Cornell University Press, Ithaca, N.Y., 1979.

Deepti, R. Haritash and Tej Verma: *Promoting Community Nutrition Through Participatory Monitoring and Evaluation*, Advance, 2001.

Drucker, Peter: *Work and Tools*, Technology and Culture, New York, 1972.

E. Ramann: *The Evolution and Classification of Soils*, Asiatic Pub, Calcutta, 2006.

Featherly H. I.: *Taxonomic Terminology of the Higher Plants*, USA, Iowa State College Press, 1954.

Feynman, Richard: *The Feynman Lectures on Physics;* Addison Wesley, U.S.A., 1964.

Finston, H.L. and A. C. Rychtman: *A New View of Current Acid-Base Theories*, John Wiley & Sons, New York, 1983.

Flond, R.: *The British Machine-tool Industry: 1850-1914,* Cambridge University Press, New York, 1976.

Francis Durso, Patricia DeLucia: "Engineering Psychology", *The Corsini Encyclopedia of Psychology*, John Wiley and Sons, NY, 2010.

Giacovazzo, C; Monaco HL, Viterbo D, Scordari F, Gilli G, Zanotti G, and Catti M.: *Fundamentals of Crystallography*, Oxford University Press, Oxford, 1992.

Gowdy, J., *Coevolution Economics: The Economy, Society and the Environment*, Kluwer, Boston, 1994.

Green, D.J.; Hannink, R.; Swain, M.V.: *Transformation Toughening of Ceramics*, CRC Press, Boca Raton, 1989.

Ihde, Aaron John: *The Development of Modern Chemistry*, Courier Dover Publications, 1984.

Johannes G. de Vries, Cornelis J. Elsevier: *The Handbook of Homogeneous Hydrogenation,* Wiley-VCH, Weinheim, 2007.

John A, Marcell, *Handbook of Small Business Valuation Formulas and Rules of Thumb,* Valuation Press, 1993.

John D. Roberts, Marjorie C. Caserio: *Basic Principles of Organic Chemistry*, W. A. Benjamin, Inc., 1964.

Jones, Denny: *Principles and Prevention of Corrosion,* Upper Saddle River, Prentice Hall, New Jersey, 1996.

Kittel and Kroemer: *Thermal Physics*, W. H. Freeman, New York, 1980.

Lata Bhattacharya: *Biochemistry of Nutrition*, Discovery, Delhi, 2010.

Leavitt, H.J.: *Applied Organizational Change in Industry: Structural, Technological and Humanistic Approaches*, Rand McNally, Chicago, 1965.

Lovesey, S. W.: *Theory of Neutron Scattering from Condensed Matter; Neutron Scattering*, Clarendon Press, Oxford, 1984.

March, Jerry: *Advanced Organic Chemistry: Reactions, Mechanisms, and Structure,* Wiley, New York, 1985.

Mayr, O.: *The Origins of Feedback Control*, MIT Press, Cambridge, 1970.

McCrum, N. G.; Buckley, C. P.; Bucknall, C. B.: *Principles of Polymer Engineering*, Oxford University Press, New York, 1997.

McHugh, Jeanne: *Alexander Holley and the Makers of Steel,* Johns Hopkins University Press, Baltimore, 1980.

Misner, Thorne, Wheeler: *Gravitation*, W. H. Freeman, San Francisco, 1973.

Nieman, Timothy A.; Skoog, Douglas A.; Holler, F. James: *Principles of Instrumental Analysis*, Pacific Grove, CA: Brooks/Cole, 1998.

Noble, David F.: *America by Design: Science, Technology, and the Rise of Corporate Capitalism*, Knopf, New York, 1977.

Ojovan, M.I., W.E. Lee: *New Developments in Glassy Nuclear Wasteforms*, Nova Science Publishers, New York, 2007.

Pacey, A.: *The Maze of Ingenuity*, MIT Press, Cambridge, MA, 1974.

Painter, Paul C.; Coleman, Michael M.: *Fundamentals of Polymer Science: An Introductory Text*, Technomic Pub. Co., Lancaster, PA., 1997.

Partington, J.R.: *A Short History of Chemistry*, New York: Dover Publications, 1937.

Rolt, L. T. C.: *A Short History of Machine Tools*, MIT Press, Cambridge, 1965.

Rubinstein, Michael; Colby, Ralph H.: *Polymer Physics*, Oxford University Press, Oxford; New York, 2003.

Schaefer, Henry F. III: *Quantum Chemistry*, Clarendon Press, Oxford, 1984.

Sharma, Pradeep : *Biochemistry and Organisation of Cells*, RBSA Pub, Delhi, 2006.

Shreve, R. Norris, and Joseph A. Brink Jr.: *The Chemical Process Industries*, McGraw Hill, New York, 1977.

Singer, C., Holy, E. J., and Holmyard, E. J., and Hall, A. R.: *A History of Technology*, Oxford University Press, Oxford, 1954.

Skoog, D.A.; West, D.M.; Holler, F.J.: *Fundamentals of Analytical Chemistry*, Saunders College Publishing, New York, 1988.

Smith, Crosbie: *The Science of Energy - a Cultural History of Energy Physics in Victorian Britain*, The University of Chicago Press, Chicago, 1998.

Strathern, P.: *Mendeleyev's Dream – the Quest for the Elements,* Berkley Books, New York, 2000.

Streitwieser, A.; Brauman J. I. and Coulson C. A.: *Supplementary Tables of Molecular Orbital Calculations*, Pergamon Press, Oxford, 1965.

Swarnim, K. : *A Textbook of Biochemistry and Microbiology*, Surendra Pub, Delhi, 2010.

Taylor, Frederick: The Principles of Scientific Management, Harper & Row, New York, 1911.

Thompson, James D.: Organizations in Action, MacGraw-Hill, New York, 1969.

Turner, W.: *Introduction to Industrial and Systems Engineering*, Prentice Hall, NY, 1992.

Urton, Gary: *The Social Life of Numbers*, Austin, University of Texas Press, 1997.

Vanangamudi, K.: *Principles and Methods of Plant Breeding*, International Book, Delhi, 2005.

Whealy K.: *The Garden Seed Inventory*, Decorah, Seed Saver Publications, 1988.

White, G.F.: *Natural Hazards: Local, National, Global*, Oxford University Press, New York, 1974.

Wickens, C. and Hollands, J.: *Industrial and Agriculture Performance*, Prentice Hall, NY, 1996.

Woodbury, R. S.: *Studies in the History of Machine Tools*, MIT Press, Cambridge, 1972.

Young, R.A.: *The Rietveld Method*, Oxford University Press & International Union of Crystallography, Oxford, 1993.

Index

F

G

H

I

L

M

N

O

P

Q

S

T

U

V

W

❑❑❑